Artificial Intelligence-Powered Electrical Machine Design

Artificial intelligence (AI) is no longer just a futuristic concept—it is transforming the way we design, build, and manage electrical machines and intelligent systems today. This groundbreaking volume showcases the convergence of AI, machine learning, and modern engineering, offering a window into the technologies shaping tomorrow's mobility, energy, and digital ecosystems.

From intelligent safety in electric vehicles and predictive maintenance of machines to additive manufacturing, smart disaster management, and Industry 5.0 integration, each chapter presents state-of-the-art research with practical applications. Readers will discover how advanced algorithms enhance machine design, optimize charging station placement, secure cloud pipelines, and redefine data-driven decision-making. This book not only delivers technical depth but also addresses ethical and societal questions, ensuring a holistic understanding of AI's role in the modern world.

Ideal for engineers, researchers, and technology leaders, this book provides both inspiration and actionable insights. Academics and graduate students will find it a rich resource for advancing research, while professionals in mobility, energy, and Internet of Things industries can directly apply its ideas to real-world innovation. For anyone eager to harness the power of AI in engineering and beyond, this volume is an indispensable guide.

Artificial Intelligence-Powered Electrical Machine Design

Optimization, Control, and Innovation

Edited by

Krishan Arora, Himanshu Sharma, and Dwarkadas Prahladadas Kothari

CRC Press
Taylor & Francis Group
Boca Raton London New York

CRC Press is an imprint of the
Taylor & Francis Group, an **informa** business

Cover: Shutterstock.com

First edition published 2026
by CRC Press
2385 NW Executive Center Drive, Suite 320, Boca Raton FL 33431

and by CRC Press
4 Park Square, Milton Park, Abingdon, Oxon, OX14 4RN

CRC Press is an imprint of Taylor & Francis Group, LLC

ISBN: 9781041076551 (hbk)
ISBN: 9781041076568 (pbk)
ISBN: 9781003641582 (ebk)

DOI: 10.1201/9781003641582

Typeset in Sabon
by codeMantra

Contents

Preface

The field of electrical machine design is undergoing a transformative evolution, driven by rapid advancements in computational intelligence, data analytics, and automation technologies. At the forefront of this revolution is **Artificial Intelligence (AI)**—a game-changing enabler that is redefining traditional paradigms of design, control, diagnostics, and performance optimization in electrical machines. The convergence of AI techniques such as machine learning, deep learning, evolutionary algorithms, and intelligent control with electrical engineering is no longer a futuristic vision—it is today's reality, reshaping industries, improving efficiency, and opening avenues for innovation that were once unimaginable.

This book, *Artificial Intelligence-Powered Electrical Machine Design: Optimization, Control, and Innovation,* has been conceived to serve as a comprehensive resource for researchers, practitioners, and students engaged in this exciting interdisciplinary domain. This book brings together a diverse collection of contributions from leading experts and innovators who explore how AI is being seamlessly integrated into various aspects of electrical machine systems—from conceptual design and modeling to fault detection, control strategies, condition monitoring, and performance optimization.

Each chapter presents cutting-edge methodologies, real-world case studies, and insights into future directions, providing readers with both the theoretical underpinnings and practical applications of AI in machine design. Emphasis has been placed on optimization techniques, data-driven predictive control, energy efficiency, smart diagnostics, and the role of AI in sustainable and intelligent power systems.

We believe this book will be an asset for engineers, academicians, R&D professionals, and postgraduate students who are exploring or actively working in the intersections of AI and electrical engineering. As the demand for intelligent, adaptive, and energy-efficient electrical systems continues to grow, we hope this volume will inspire new ideas, foster innovation, and contribute to the ongoing advancement of smart technologies in electrical machine design.

We sincerely thank all contributing authors, reviewers, and the publishing team for their valuable inputs and commitment in bringing this collaborative work to fruition.

Editor biographies

Dr. Krishan Arora is associated with Lovely Professional University as a Professor with more than 17 years of experience in academics and research. He is currently the Head of Department of Power Systems in the School of Electronics and Electrical Engineering, Lovely Professional University. He has completed his Ph.D. in Electrical Engineering from MM (Deemed to be) University, Ambala. His research interests include electrical machines, artificial intelligence, load frequency control, electric vehicles, waste management, renewable energy, Internet of Things, and modernization of smart grids. He has published more than 110 research papers in reputed international journals. He has organized several workshops, summer internships, international conferences, special sessions, and expert lectures. He has also published ten Indian patents and one copyright. He has supervised one Ph.D. scholar, ten postgraduate scholars, and more than 30 undergraduate students' projects. He has attended/participated in 24 national/international online webinars. He has edited fifteen books in different areas of electronics and electrical engineering.

Dr. Himanshu Sharma has completed his Ph.D. in Power Electronics from Punjab Engineering College, Chandigarh. He did his M.Tech. in Electrical Engineering from Punjab Engineering College, Chandigarh. He was associated with Lovely Professional University in the Division of Research & Development with more than eight years of experience in intellectual property rights and academics. He is currently deputed as Dean R&D at G.H. Raisoni College of Engineering and Management, Nagpur, since December 2024. He has published more than 30 research papers in reputed international journals. He has organized several workshops, summer internships, international conferences, and expert lectures for students. He has supervised more than five Ph.D. students and more than 15 UG/PG students. He has attended/participated in 20 national/international online webinars. His area of expertise includes power electronics, waste management, machine learning, and optimization techniques. He has taught various courses at UG and PG levels like power electronics, nonconventional energy sources, electric drives, wireless communication, Internet of Things, etc.

Dr. Dwarkadas Prahladadas Kothari is a renowned Indian academician and researcher in the field of electrical engineering, with a distinguished career spanning over five decades. He is best known for his significant contributions to power systems, energy management, and electrical machines. A former Director of IIT Delhi and Vice Chancellor of VIT University, Vellore, Dr. Kothari has held various prestigious academic and administrative positions, including Principal of Visvesvaraya National Institute of Technology (VNIT) Nagpur and Head of the Centre for Energy Studies at IIT Delhi. He earned his B.E., M.E., and Ph.D. degrees from BITS Pilani. Widely respected in academia, he has authored over 60 books and published more than 800 research papers in national and international journals and conferences. His work on optimization techniques in power systems, energy conservation, and smart grid technologies has had a lasting impact. Dr. Kothari is a fellow of several professional bodies including IEEE, INAE, and IE(I) and has received numerous accolades such as the Lifetime Achievement Award, UGC Swami Pranavananda Saraswati Award, and the Khosla National Award. His legacy includes mentoring over 50 Ph.D. scholars and serving as a source of inspiration to generations of engineers and researchers in India and abroad.

Contributors

Andi Adawiah
Department of Management
Universitas Lamappapoleonro
Soppeng, Indonesia

B. Amutha
Department of Computing Technology
SRM Institute of Science and Technology
Chennai, India

Jayasri Angara
Department of Computer Science Engineering
GITAM University
Visakhapatnam, India

Krishan Arora
School of Electronics and Electrical Engineering
Lovely Professional University
Punjab, India

Ranjit Kumar Bindal
Electrical Engineering Department
Chandigarh University
Mohali, India

Neeraj Chandnani
KLEF (Koneru Lakshmaiah Education Foundation)
Deemed to be University, Vaddeswaram, Guntur
Andhra Pradesh, India

Harpreet Kaur Channi
Department of Electrical Engineering
Guru Nanak Dev Engineering College
Ludhiana, India

Anita Gehlot
Division of Research and Innovation
Uttaranchal Institute of Technology Uttaranchal University
Dehradun, India

Deepika Ghai
School of Electronics and Electrical Engineering
Lovely Professional University
Phagwara, India

Amit Goswami
Source Infotech
Edison, New Jersey

S. B. Goyal
Faculty of Information Technology
City University
Petaling Jaya, Malaysia

Mahim Raj Gupta
Western Digital (SanDisk Corporation)
San Jose, California

Vaibhav Jadon
Electrical Engineering Department
Chandigarh University
Mohali, India

Pawandeep Kaur
School of Electronics and Electrical Engineering
Lovely Professional University
Phagwara, India

Pankaj Kumar Keshri
School of Electronics and Electrical Engineering
Lovely Professional University
Phagwara, India

Ashish Khaira
Department of Mechanical Engineering
School of Engineering and Technology, Adamas University
Kolkata, West Bengal, India

A. Mohan Kumar
Department of Mechanical Engineering
Kongu Engineering College
Erode, India

Prashant Kumar
Electrical Engineering Department
Chandigarh University
Mohali, India

Sandi Lubis
Department of Government Affairs and Administration
Universitas Muhammadiyah Yogyakarta
Yogyakarta, Indonesia

Praveen Kumar Malik
School of Electronics and Electrical Engineering
Lovely Professional University
Phagwara, India

Halimah Abdul Manaf
UUM College of Law, Government, and International Studies
University Utara Malaysia
Kedah, Malaysia

Chirag Mavani
DXC Technology
Ashburn, Virginia

Eliza Meiyani
Department of Government Affairs and Administration
Universitas Muhammadiyah Yogyakarta
Yogyakarta, Indonesia

Hirenkumar Mistry
Sr Linux Administrator & Cloud Engineer
Zenosys
Kendall Park, New Jersey

Dyah Mutiarin
Department of Government Affairs and Administration,
Universitas Muhammadiyah Yogyakarta
Yogyakarta, Indonesia

Achmad Nurmandi
Department of Government Affairs and Administration
Universitas Muhammadiyah Yogyakarta
Yogyakarta, Indonesia

Fraiz Parveen
Division of Research and Innovation
Uttaranchal University
Dehradun, India

Satya Ravi Teja Peddada
Department of Electrical and Computer Engineering
University of Illinois Urbana-Champaign
Urbana, Illinois

Andi Luhur Prianto
Department of Government Studies
Universitas Muhammadiyah Makassar
Makassar, Indonesia

Ulung Pribadi
Department of Government Affairs and Administration
Universitas Muhammadiyah Yogyakarta
Yogyakarta, Indonesia

Titin Purwaningsih
Department of Government Affairs and Administration
Universitas Muhammadiyah Yogyakarta
Yogyakarta, Indonesia

K. P. Ranjusha
Department of Computing Technology
SRM Institute of Science and Technology
Chennai, India

Ramandeep Sandhu
School of Computer Science & Engineering
Lovely Professional University
Phagwara, India

Praseeda Saripalle
Department of Computer Science
Gayatri Vidya Parishad College of Engineering for Women
Visakhapatnam, India

Ravi Shankar Saripalle
Department of Entrepreneurship
GITAM University
Visakhapatnam, India

Sumit Sunil Shelar
School of Electronics and Electrical Engineering
Lovely Professional University
Phagwara, India

Rajesh Singh
Division of Research and Innovation
Uttaranchal Institute of Technology
Uttaranchal University
Dehradun, India

N. S. Sivakumar
Department of Engineering
Nusa Putra University
Sukabumi, Indonesia

Imron Sohsan
Faculty of Humanities and Social Science
Khon Kaen University
Khon Kaen City, Thailand

P. Sureshkumar
Department of automobile Engineering
JCT College of Engineering & Technology
Coimbatore, India

Kumud Tiwari
School of Electronics and Electrical Engineering
Lovely Professional University
Phagwara, India

Muhammad Younus
Department of Government Affairs and Administration
Universitas Muhammadiyah Yogyakarta
Yogyakarta, Indonesia

Chapter 1

Integration of intelligent safety systems in electric vehicle drive machines

Krishan Arora

1.1 INTRODUCTION

Transportation is a vital part of human survival. It makes our life easier. In today's world, we are so dependent on vehicles that we cannot sustain ourselves without them. Even just for short distances, we use vehicles. We are widely dependent on conventional internal combustion engine automobiles for our travel. But unfortunately, our conventional internal combustion engine automobiles use nonrenewable resources. Renewable resources are very sparse in our environment. Thus, extensive use of it is not a sustainable option. Moreover, conventional combustion engine vehicles produce a significant number of pollutants in the environment. This is why EVs (electric vehicles) are the future for vehicles.

EVs are the most sustainable and clean form of automobile. They use electrical energy (EE) to rotate the wheels. Therefore, they don't produce pollutants and reduce the emission of greenhouse gases. Thus, they are more sustainable than existing conventional combustion vehicles. But unfortunately, efficiency and safety are the major concerns in EVs.

EVs take a lot of time to charge. Hence, our system must waste very less energy. Therefore, with our system, we aim to make our EV more efficient by utilizing the energy that is wasted while traveling and braking. So, we are using this energy to run the car and the safety system in our car. Hence, we can make our EV more efficient. So, we are using two methods for it:

- **Regenerative Braking System**: Our regenerative braking system recovers up to 16%–70% depending on the driver's style of driving whether they brake gradually or severely and use it to recharge the battery of the vehicle and power up the vehicle safety system.
- **Dynamo as a Generator**: In the dynamo as a generator, our main objective is to store the inertial energy when the user does not apply acceleration ($a = 0\,m/s^2$). For instance, during downhill, the user does not have to apply acceleration since no propulsion power is required

DOI: 10.1201/9781003641582-1

from the energy source. Thus, at the same instance, we can harvest the energy from the rotating wheel and utilize it in our battery and smart safety system.

The safety of occupants is a major concern for any automobile. Thus, with our system, we aim to provide safety before and after the accident. This system includes the following:

- **Colloidal Avoidance System**: In the colloidal avoidance system (CAS), we are detecting the vehicles in front of us so that it will inform the driver to slow down and avoid the accident. Moreover, we are also adding alcohol detection, which will not allow a drunken person to start the vehicle.
- **Colloidal Detection System (CDS)**: In the CDS, we are using Global System for Mobile Communication (GSM) technology to alert the driver's well-wishers about the accident by sending the vehicle's accident location calculated by the Global Positioning System (GPS) module. With the help of the driver's location, they can be saved and appropriately helped (Figure 1.1).

1.2 LITERATURE SURVEY

1. **Powering the Electric Cars with Dynamos**
   In this paper, the author suggests powering up the electric car with dynamos. The system uses the technique that places one dynamo in each wheel, to produce a charge through the rotatory motion of wheels, and these charges are stored in a separate battery that gives the supply for the motor. This energy can later be used for other

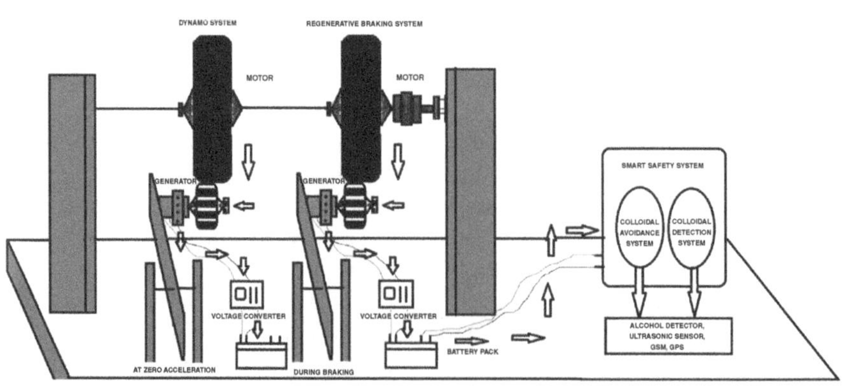

Figure 1.1 Modeling idea of the project.

emergency purposes, and this process is cyclic. Thus, making the car run for longer distances [1].

2. **Regenerative Braking Control Method Based on Predictive Optimization for Four-Wheel Drive Pure EV**

   In the following research paper, the author gives the data showing that 50% of the energy of EVs during the conventional braking system is lost in the form of heat in urban conditions. The author suggests that the regenerative braking system should be used to overcome this problem and utilize the energy in the best way [2].

3. **Research on Regenerative Braking Systems: A Review**

   In this research paper, the author talks about the environmental problems and energy crises due to conventional internal combustion engines and shifting toward EVs installed with regenerative braking systems. EVs contribute to decreasing the pollution level, and the regenerative braking system retains the energy losses and uses them for various purposes. The author suggests various types of regenerative braking systems, i.e., series regenerative braking systems and parallel regenerative braking systems [3].

4. **Portable Alcohol Detection System for Driver Monitoring**:

   In this paper, the author suggests using exhaled breath to analyze the presence of alcohol. The whole system includes the exhaled air gas sensor and mobile phone, which helps communicate various data in the cloud. This system can easily monitor the alcohol content if the driver is from anywhere. The author uses four sensors, i.e., one water vapor sensor and three semiconductor sensors like ethanol, acetaldehyde, and hydrogen, and gives the data to the cloud to detect alcohol [4].

5. **Alcohol Detection for Car Locking System**

   In this study, the author proposes a locking system in the car, which will not start until passing the alcohol detection mechanism. The author uses Arduino Uno along with an MQ3 alcohol detector, where MQ3 is used to detect alcohol and Arduino Uno controls the locking of wheel accordingly [5].

6. **Design of Handheld Positioning Tracker Based on GPS/GSM**

   In this paper, the author suggests the use of GPS navigation and GSM wireless data transmission for developing a positioning tracker. The author utilizes the STC12C5A60S2 to receive the GPS positions, and the received data is henceforth transmitted to the user's mobile using the GSM transmission. By this method, the user can anytime request the location from the system and it will provide the position [6].

7. **Automobile Antitheft System Based on GSM and GPS Modules**

   In this study, the author proposes an antitheft tool using the GSM and GPS modules. Here, the author had developed a system that has a C8051F120 chip, which monitors the system using a vibration sensor.

We are using a GPS that will collect the position of the system. By using the GSM transmission, we will transmit this location to the user's mobile [7].

1.3 DESIGN METHODOLOGY

In our model, we are basically harvesting the wasted energy from the regenerative braking system during braking and the dynamo as a generator during zero acceleration to charge the battery and smart safety system (including the CAS and CDS) as shown in Figure 1.2.

Since the energy generated from the regenerative braking system and dynamo as a generator system is high, it is used for our safety system. We pass it through a buck converter, which converts the generated energy to 5 V. Thus, it makes it fit to run Arduino Uno and other sensors. Therefore, we are increasing the efficiency of the vehicle by conserving wasted energy and utilizing it to charge the battery and run our safety system (CAS and CDS).

1.4 COMPONENT DESCRIPTION

1. **Controller**: It helps maintain the stability of voltage as there will not be a continuous flow of voltage while moving the wheel (Figure 1.3).
2. **Motor/Generator**: A motor is a device that basically provides the rotation by converting EE into mechanical energy. In this prototype, a motor is used to run the wheel of a vehicle (Figure 1.4).
3. **Brake Drum**: A brake drum is a part of the drum braking system used in cars to reduce speed or stop the car. It is a steel or cast-iron

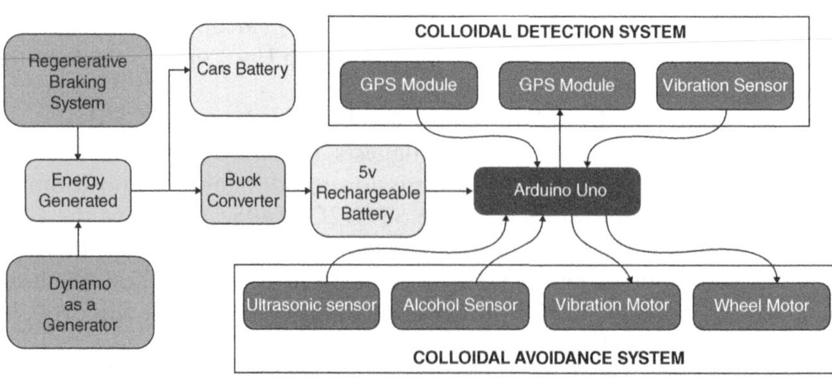

Figure 1.2 Block diagram of the proposed model.

Figure 1.3 Controller.

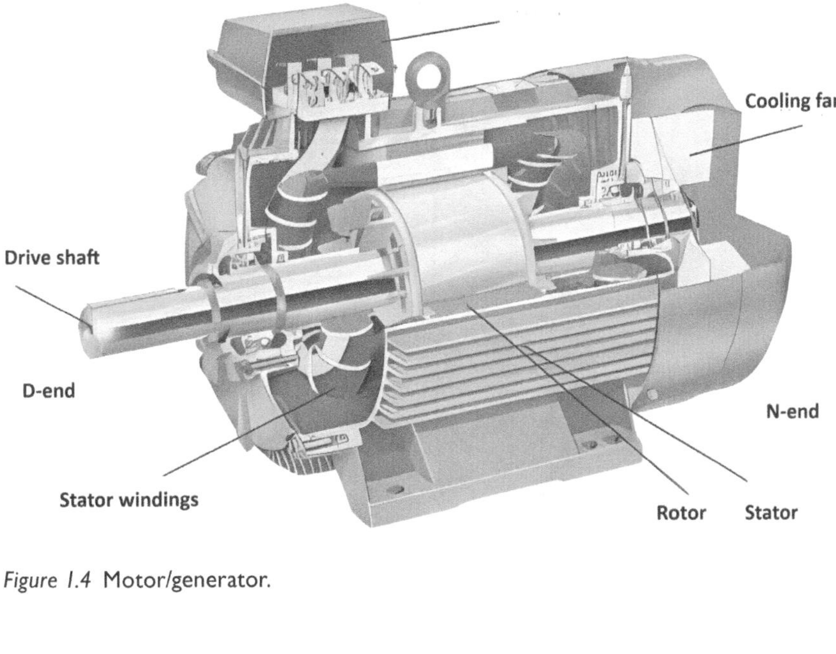

Figure 1.4 Motor/generator.

drum with a circular shape that is mounted on the axle of a vehicle (Figure 1.5).

4. **Linking Mechanism**: It is a mechanism to link the brake pedal to the DC motor/generator (Figure 1.6).
5. **Friction Lining:** It is a component used in braking systems to generate friction, slow down, or stop a moving vehicle. It is known as brake lining. Typically, a composite material is used to create it.

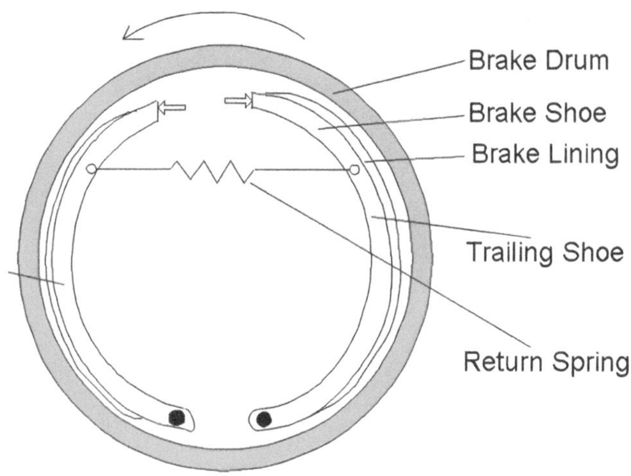

Figure 1.5 Brake drum.

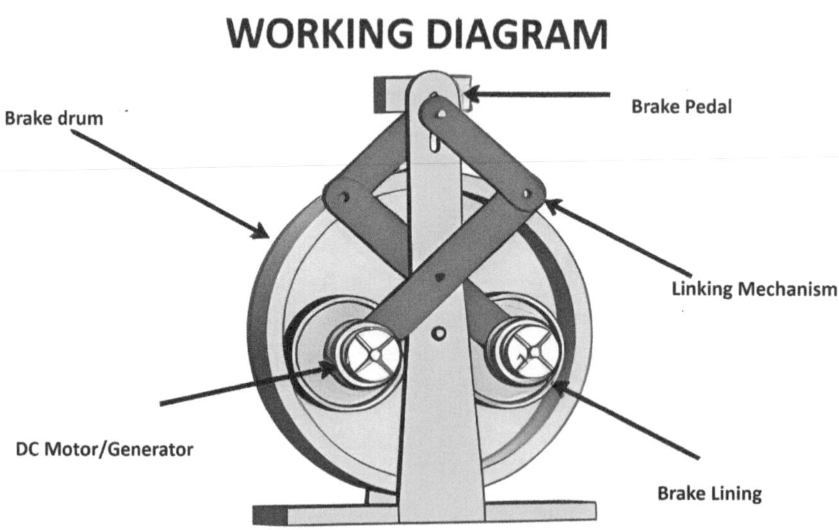

Figure 1.6 Linking mechanism.

6. **Dynamo**: A dynamo is an electrical device that generates DC using the commutator. This prototype uses it to create energy while in contact with the wheels and transfer it to the battery and the vehicle's safety sensors (Figure 1.7).
7. **Wheels**: (Figure 1.8)

Figure 1.7 Dynamo.

Figure 1.8 Wheels.

8. **Bridge Rectifier**: It is an electrical device that is used to convert AC to DC. This type of system consists of four diodes connected in a bridge configuration.
9. **DC–DC Boost Converter**: It is an electrical device that steps down the voltage of the system (Figure 1.9).
10. **Battery**: A battery is a power source that is used to provide energy to the system (Figure 1.10).
11. **Arduino Uno**: Arduino Uno is an ATmega328P microcontroller development board consisting of various features. It was designed to make electronics more accessible to hobbyists, artists, and designers, with the aim of allowing anyone to build interactive projects with minimal knowledge of electronics and programming.

    The Arduino Uno board features 14 digital input/output pins, six analog inputs, a 16 MHz quartz crystal, a USB connection for programming and power, an In-Circuit Serial Programming header, and a reset button. Arduino uses a USB or an external supply to power itself (Figure 1.11).
12. **Ultrasonic Sensor**: An ultrasonic sensor is a device that is used to measure the distance of any object from the sensor. It uses ultrasonic waves for it. It has a transmitter that releases 40 kHz waves upon triggering with 10 μs signal in the trigger pin. These waves hit the object and return and are collected by a receiver, which generates a signal

Figure 1.9 DC–DC boost converter.

Figure 1.10 Battery.

Figure 1.11 Arduino Uno.

equivalent to the time it took to hit the object and sends it to the echo pin (Figure 1.12).

13. **Microvibration Motor**: Small, highly torqued motors called "micro-vibration motors" are used to create vibration. The stator houses a collection of coils that generate a magnetic field, whereas the rotor is often a small eccentric weight coupled to a shaft. The magnetic field causes the rotor to revolve quickly when an electrical current is delivered to the coils, causing vibrations. Mobile gadgets, game controllers, and other electrical devices that require vibration feedback all frequently use them. In our prototype, it is used on the steering wheel to create a vibration that feels to the driver at any moment (Figure 1.13).
14. **MQ-3 Gas Sensor**: It is a dual-function module with GSM and GPS. Based on the most recent GSM/GPS module from SIMCOM, the SIM808, it utilizes GPS technology for satellite navigation and supports GSM/GPRS Quad-Band networks. It has a very long standby life and a charging circuit for Li-Ion batteries built right in. This makes it perfect for projects that call for rechargeable Li-Ion batteries. It contains 66 acquisition and 22 tracking receiver channels, which contribute to its high GPS reception sensitivity (Figure 1.14).

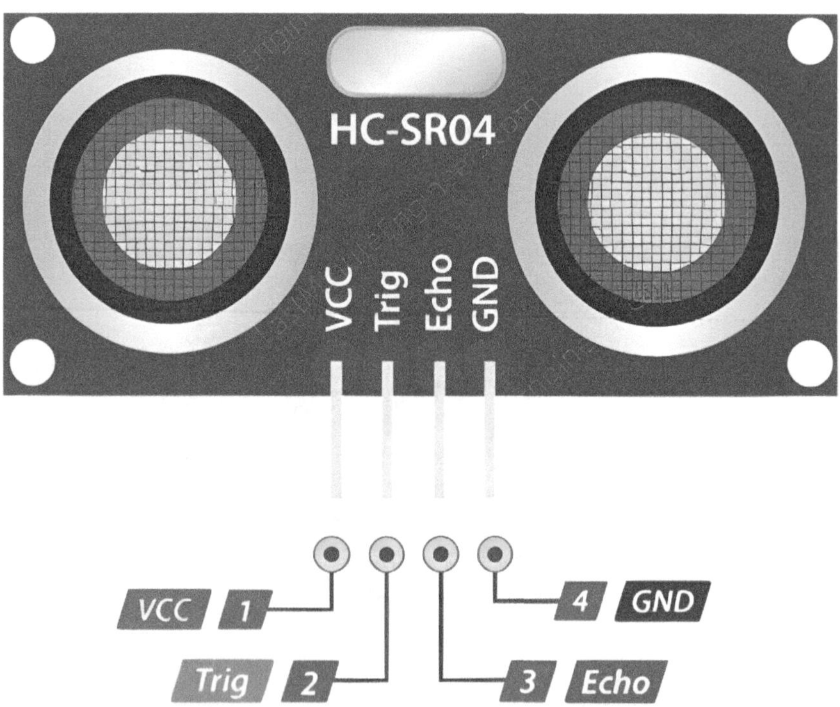

Figure 1.12 Ultrasonic sensors.

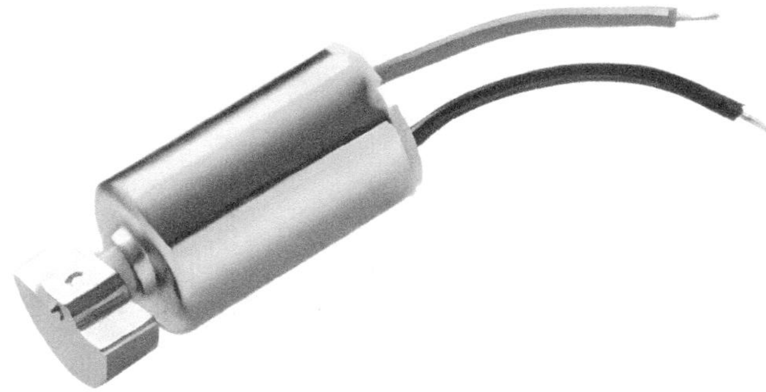

Figure 1.13 Microvibration motor.

Figure 1.14 MQ-3 gas sensor.

15. **GSM Module SIM808**: It is a dual-function module with GSM and GPS. It utilizes GPS technology for satellite navigation and supports GSM/GPRS Quad-Band networks. It has a very long standby life and a charging circuit for Li-Ion batteries built right in. This makes it perfect for projects that call for rechargeable Li-Ion batteries. It contains 66 acquisition and 22 tracking receiver channels, which contribute to its high GPS reception sensitivity (Figure 1.15).
16. **GPS Module Neo-6M**: It is an electronic device that is used to find the location of the system. It can tell you about various geographical factors like latitude, longitude, and height (Figure 1.16).
17. **Vibration Sensor sw-420**: It is a nondirectional vibration sensor with excellent sensitivity. It gives a high signal when the vibration value is high; otherwise, it will give a low signal (Figure 1.17).

1.5 FULL PROTOTYPE

See Figure 1.18.

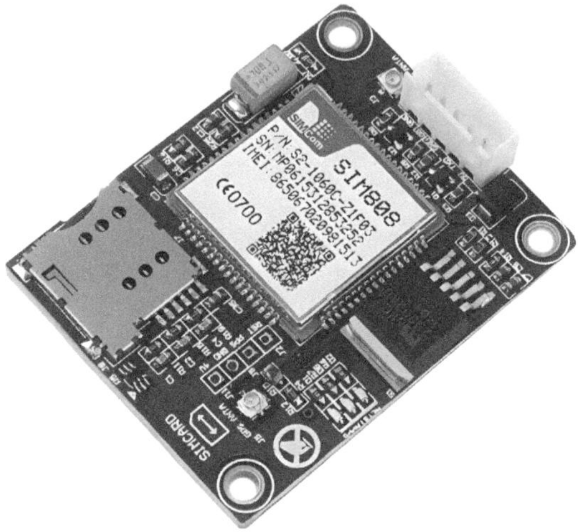

Figure 1.15 GSM module SIM808.

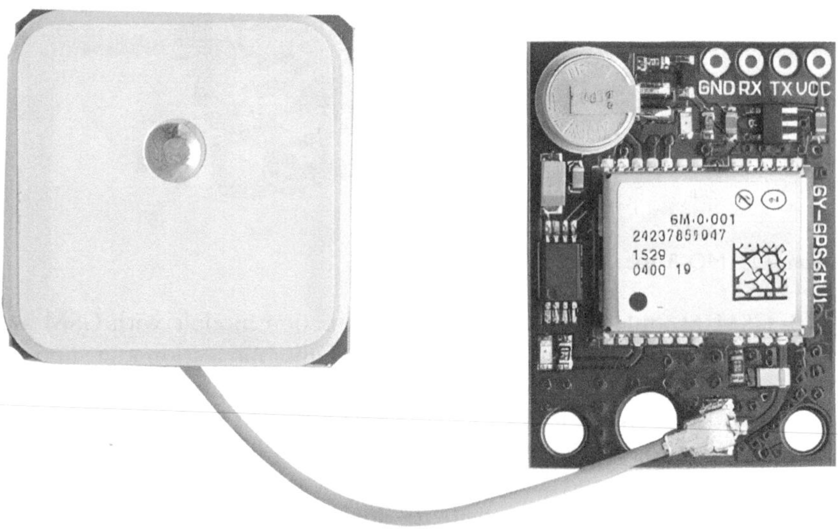

Figure 1.16 GPS module Neo-6M.

1.6 DETAILED DESCRIPTION

1.6.1 Regenerative braking system

To improve the efficiency of EVs, we are going to install a regenerative braking system. Research tells us that during conventional braking systems or deceleration, EVs waste almost 50% of their energy due to various

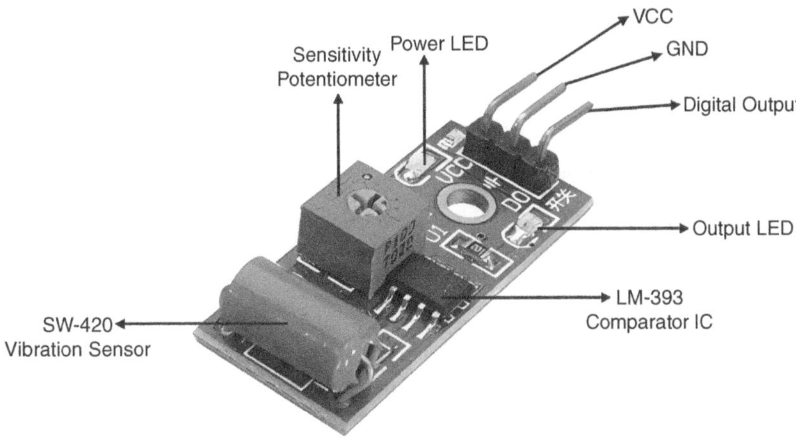

Figure 1.17 Vibration sensor sw-420.

Figure 1.18 Full prototype.

losses in urban conditions. It recovers up to 16%–70% of the kinetic energy (KE) which would otherwise be wasted as heat and converts it into EE and recharge the battery of the vehicle. So, the mileage of the vehicle will increase [8].

1.6.1.1 Components

1. Battery
2. Controller

3. Motor/generator
4. Brake drum
5. Linking mechanism
6. Friction lining

1.6.1.2 Block diagram

See Figure 1.19.

1.6.1.3 Working

The principle behind the regenerative braking is the law of conservation of energy, which signifies that the total energy of any remote system stays the same. We can convert them for useful purposes.

Whenever the brake is applied or the vehicle decelerates, a lot of KE is lost in the form of heat, so to reduce these energy losses, a regenerative braking system is used. It works when the brake is applied the electric motor acts as a generator and starts generating electricity, which can be stored in a battery for later use.

The power flow in EVs is from the battery to the wheels for propulsion of the vehicle (Figure 1.20).

The power flow in the regenerative braking system is from the wheels to the battery for recharging purposes (Figures 1.21 and 1.22).

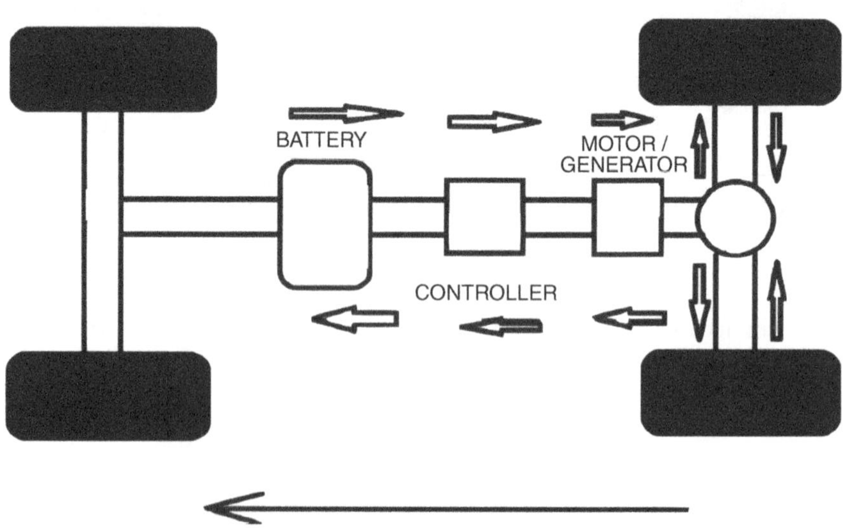

Figure 1.19 Block diagram of the regenerative braking system.

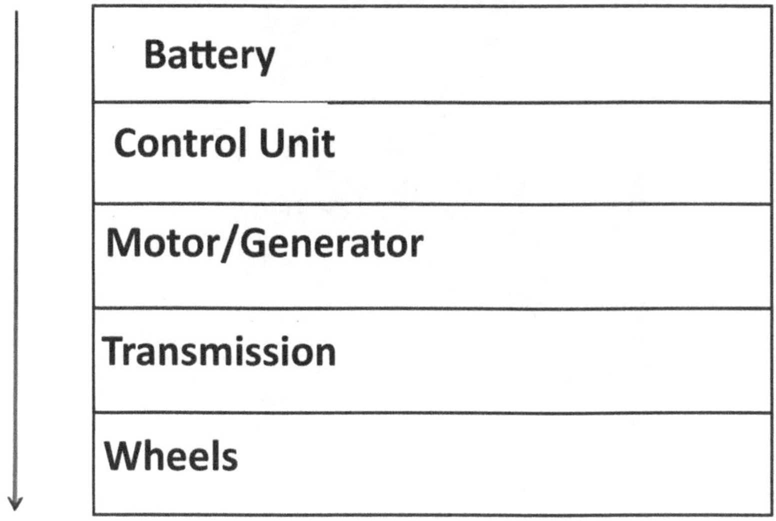

Figure 1.20 Flow chart.

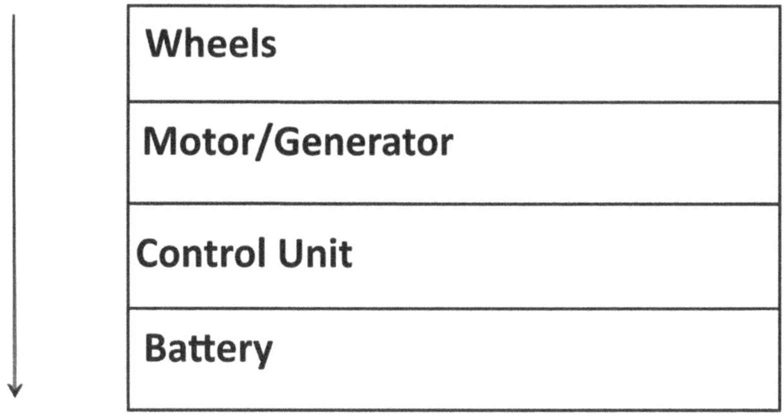

Figure 1.21 Flow chart.

1.6.2 Dynamo as a generator

Usually, a dynamo is used to distribute or to regenerate energy from wheels. But it decreases the efficiency of the system. Therefore, we are using dynamo to store the energy without hampering the efficiency of the vehicles. In our system, when the acceleration of the vehicle is zero, the dynamo's system engages with wheels and starts converting the KE of vehicle's inertial motion into EE, which will basically recharge our vehicle's battery and our safety system.

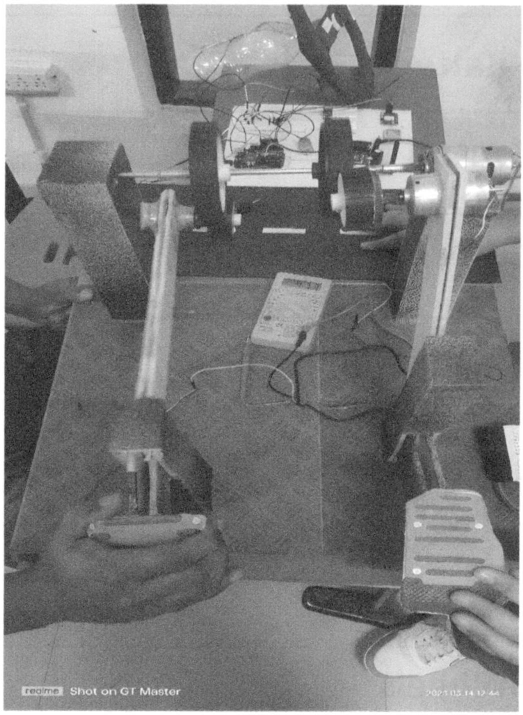

Figure 1.22 Regenerative braking system (we are getting 4.73 V after applying the brake pedal due to the regenerative braking system).

1.6.2.1 Components

1. Dynamo
2. Wheel
3. Bridge rectifier
4. DC–DC boost converter
5. Battery

1.6.2.2 Block diagram

See Figure 1.23.

1.6.2.3 Working

In this part, when the acceleration of the vehicle is zero or no propulsion energy is needed, then the dynamo system engages with the wheel to convert KE into EE due to the inertia of the motion of the wheel.

The bridge rectifier will convert the AC output of the dynamo into DC output, which is connected to the controller (DC–DC boost). This controller

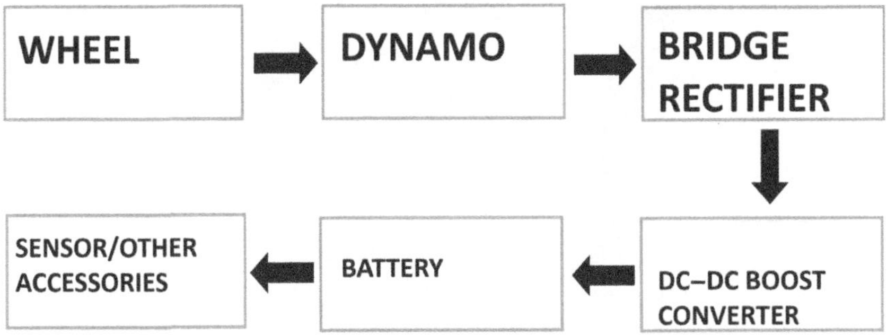

Figure 1.23 Block diagram of a dynamo as a generator.

is a mini stabilizer, and we have used it because when the vehicle starts, its speed will vary.

Therefore, the voltage of this dynamo will fluctuate, due to which we will not be able to charge the battery properly, but the controller installed in it will maintain the fluctuating volts and amperes, both will be constant. Now, these energies are fed to our battery and safety system. This energy is not feasible for our safety system as our Arduino can only stand up to 5 V.

Thus, we are installing a buck converter that converts 12 V into 5 V.

Dynamo as a generator generates 12 V, which is passed through a buck converter giving 4.66 V when the acceleration pedal is at zero or no acceleration is given (Figure 1.24).

1.6.3 Colloidal avoidance system

In the CAS, we are using an ultrasonic sensor to detect objects in front of the driver and alert the driver by vibrating the microvibration motor in the steering wheel. In the CAS, we are also installing an alcohol detection system using the MQ-3 alcohol sensor, which will stop the car if alcohol is detected.

1.6.3.1 Components

1. Arduino Uno
2. Ultrasonic sensor
3. Microvibration motor
4. MQ-3 gas sensor

1.6.3.2 Block diagram

In our CAS block diagram, we are installing an ultrasonic sensor at the frontal part of the vehicle so that it can detect objects or vehicles in front of us.

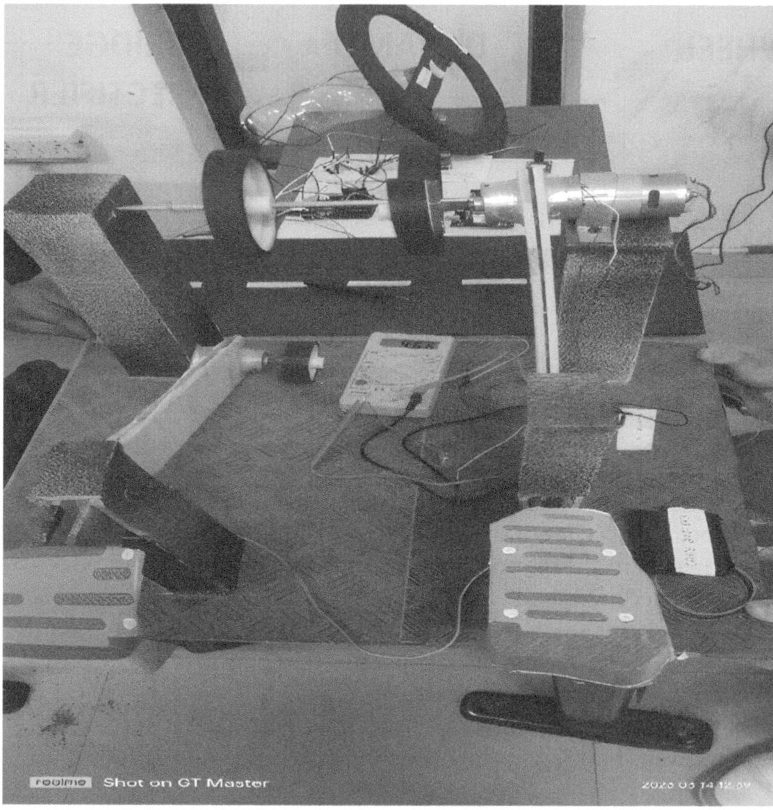

Figure 1.24 Dynamo as a generator (energy generated is 12 V, which is passed through a buckconvertergiving4.66 Vwhentheaccelerationpedalisatzeroornoacceleration is given).

We have installed a microvibration motor in the steering wheel, which will vibrate upon detecting any vehicle or object. We have also placed the MQ-3 sensor at the center of a steering wheel, which detects the alcohol and breaks the connection in the relay of wheels, thus stopping the vehicle. All these sensors and actuators are controlled by Arduino Uno (Figure 1.25).

1.6.3.3 Circuit diagram

See Figure 1.26.

1.6.3.4 Working

In the CAS, we are detecting the vehicles or people or animals in front of us so that it will inform the driver to slow down and avoid the accident.

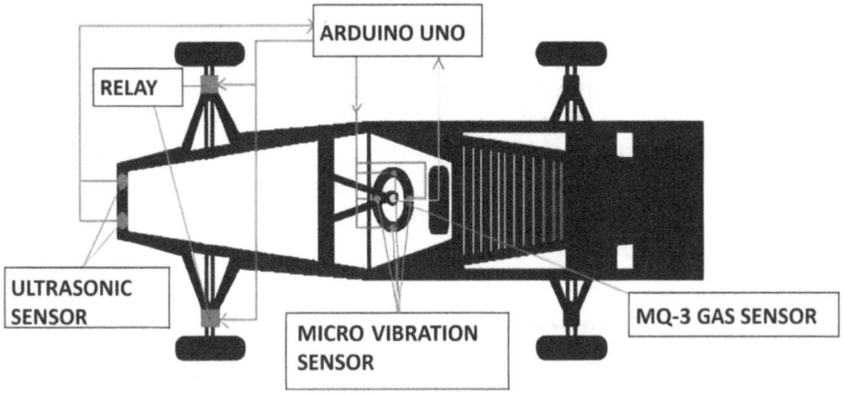

Figure 1.25 Block diagram of the colloidal avoidance system.

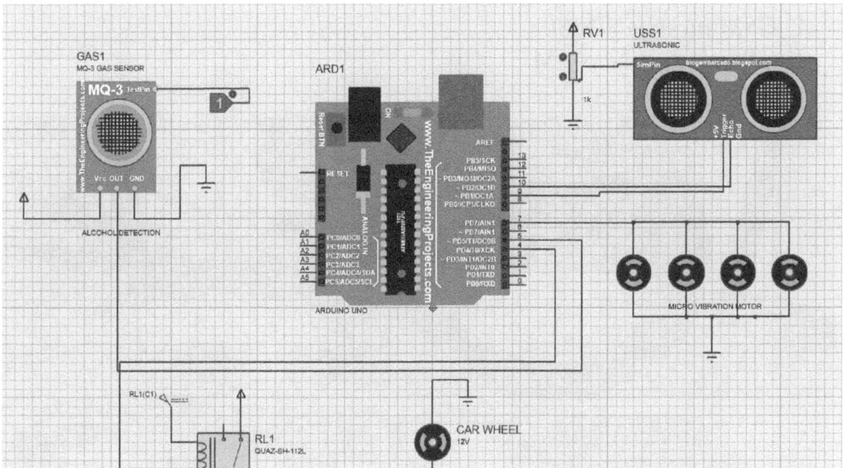

Figure 1.26 Circuit diagram of the colloidal avoidance system.

Moreover, we are also adding alcohol detection, which will not allow the drunken person to start the car.

In the model, we are using an ultrasonic sensor at the front of the vehicle. The ultrasonic sensor is a device used to measure the distance of objects. Therefore, it measures the distance of the vehicles in front of it and sends it to Arduino. Arduino will check if the distance between the vehicles is smaller than 150 cm or not. If the distance is smaller than 150 cm then, it will give power to the microvibration motor, thus causing the vibration at the steering wheel, indicating that the user needs to slow down the car.

Moreover, I am adding an MQ-3, which will sense the amount of alcohol and send that data to the Arduino Uno. If alcohol is detected, Arduino will send a high signal to the relay connected to the wheels, thus breaking the power connection of the wheels (stopping the wheels). The Arduino will send the low signal to relay, thus enabling the power of the wheels. Thus, if the user is drunk, they will not be allowed to drive and help avoid accidents (Figures 1.27 and 1.28).

1.6.4 Colloidal detection system

CDS consists of a vibration sensor with module and a GPS module, which will send the location of the vehicle after a collision and alert the user's well-wishers about the accident.

1.6.4.1 Components

1. GSM module SIM808
2. GPS module Neo-6M

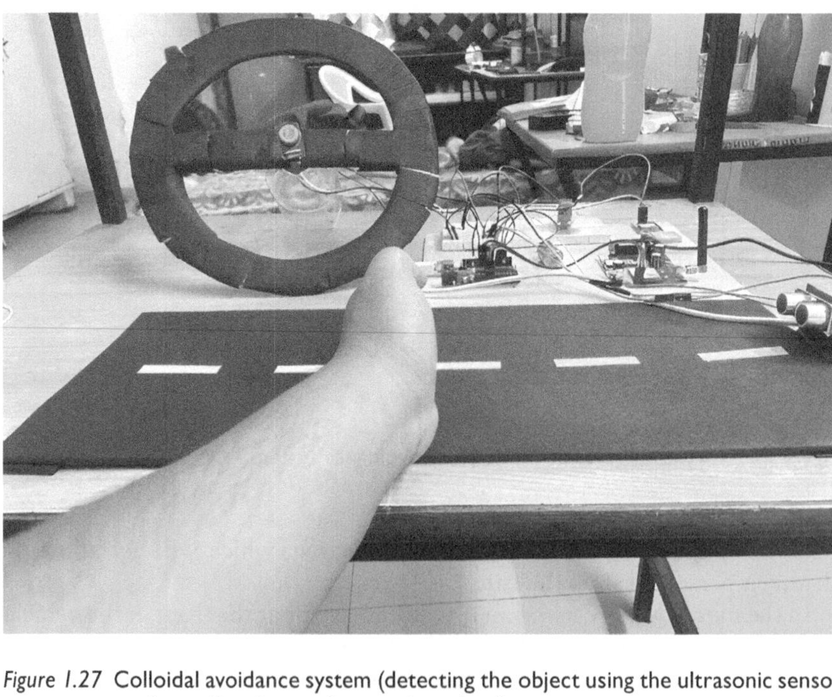

Figure 1.27 Colloidal avoidance system (detecting the object using the ultrasonic sensor to give a vibration to the steering wheel).

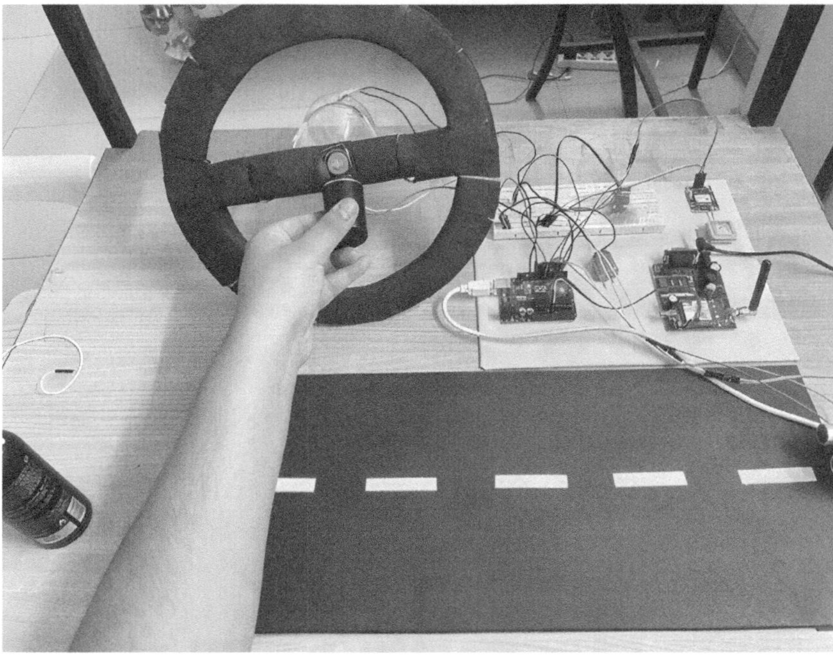

Figure 1.28 Colloidal avoidance system (detecting alcohol to stop the car from moving).

3. Arduino Uno
4. Vibration sensor sw-420

1.6.4.2 Block diagram

In our CDS model, we are installing a vibration sensor on each side of the vehicle so that if the vehicle gets collided with any side, it will detect the vibration and send a notification to the user's well-wishers. We are also placing GPS, which will locate the location of the accident and send the details of the accident to the user's well-wishers using GSM (Figure 1.29).

1.6.4.3 Circuit diagram

See Figure 1.30.

1.6.4.4 Working

In the model, we are installing the vibration sensor on each side of the vehicle. The vibration sensor detects the frequency of vibration. When the

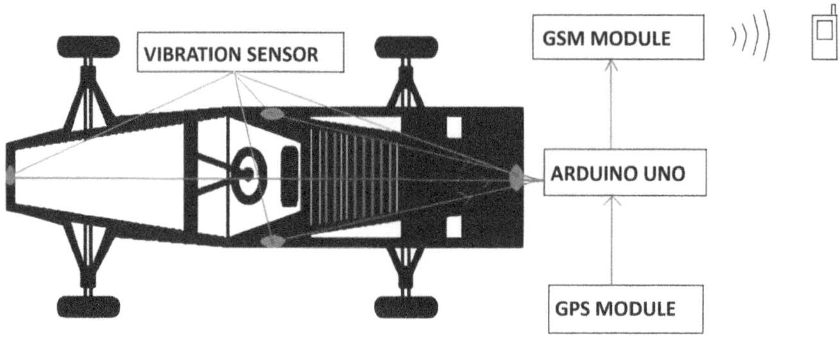

Figure 1.29 Block diagram of the colloidal detection system.

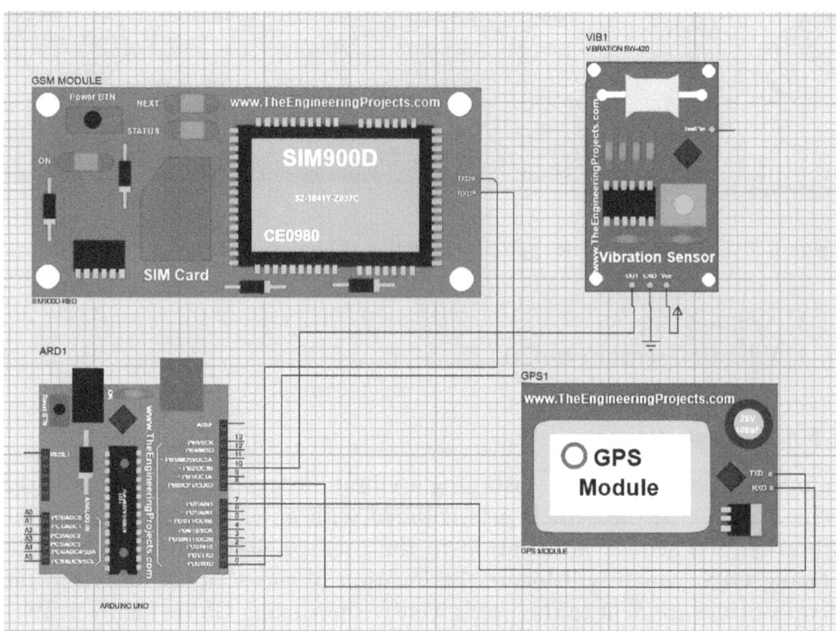

Figure 1.30 Circuit diagram of the colloidal detection system.

Figure 1.31 Colloidal detection system using GSM and GPS.

vehicle meets with an accident, vibration will be generated because of the collision, which will be measured by a vibration sensor and then sent to the Arduino. Arduino will further check whether the frequency is large or not. If the vibration frequency is large, then it will send the message to the user's mobile phone via the GSM module that "The vehicle met with an accident".

A GSM module is a device that uses mobile communication to transfer data over the network. Thus, we can send it to the mobile phone via the GSM module. Moreover, after collision, Arduino will take the location from the GPS module. The GPS module is the device that locates the current position of the device. Thus, the GPS module calibrates the location after the collision and sends the location to Arduino.

Then, Arduino sends this location to the user using the GSM module. Therefore, the user's well-wishers can be aware of the accident and rescue the person (Figure 1.31).

1.7 PROGRAM OF THE MODEL

```
const int sensor_pin=8 ;    //Alcohol sensor
const int trigpin=5;
const int echopin=4;
const int vibrations=12;
const int wheel=7;     //Car Wheels
long duration;
int distance;
#include <SoftwareSerial.h>
#include <TinyGPS.h>
int vibrationsensor =11;
float gpslat, gpslon;
TinyGPS gps;
SoftwareSerial sgps(2, 3);
SoftwareSerial sgsm(9, 10);
void setup() {
  Serial.begin(9600);
  Serial.println("System warming up!");
  delay(10000);
  pinMode(trigpin,OUTPUT);
  pinMode(echopin,INPUT);
  pinMode(vibrations,OUTPUT);
  pinMode(wheel,OUTPUT);
  pinMode(sensor_pin,INPUT);

  Serial.begin(9600);
  sgsm.begin(9600);
  sgps.begin(9600);
  pinMode(vibrationsensor , INPUT);
}

```

```

void loop() {

  // ************************ULTRASONIC SENSOR***************
  digitalWrite(trigpin,LOW);
  delayMicroseconds(2);
  digitalWrite(trigpin,HIGH);
  delayMicroseconds(10);
  digitalWrite(trigpin,LOW);

  duration=pulseIn(echopin,HIGH);
  distance=duration*0.034/2;
  Serial.print("Distance : ");
  Serial.print(distance);
  Serial.print(" cm");
  delay(100);
  if((distance<30) && (distance>2))
  {
    digitalWrite(vibrations,HIGH);   //microvibration motor on
    delay(300);
  }
  else
  {
    digitalWrite(vibrations,LOW);   //microvibration motor off
  }

  //************************Alcohol Detection***************
  int sensor_value = digitalRead(sensor_pin);
  Serial.print("  |  Digital Output: ");
  Serial.print(sensor_value);
  if(sensor_value==1)
  {
    Serial.println("  |  Alcohol: Not Detected");
    digitalWrite(wheel, HIGH);
    delay(100);
  }
  else
  {
    Serial.println("  |  Alcohol: Detected!");
    digitalWrite(wheel, LOW);
    delay(100);
  }
```

```

//*******************************GPS AND GSM ******:

sgps.listen();
while (sgps.available())
{
  int c = sgps.read();
  if (gps.encode(c))
  {
    gps.f_get_position(&gpslat, &gpslon);
  }
}
long measurement = vibration();
delay(500);
Serial.print("Vibration Value : ");
Serial.println(measurement);
if (measurement>250)
{
    Serial.print("Latitude : ");
    //Serial.print("31.249519");
    Serial.println(gpslat,6);
    Serial.print(" | Longitute: ");
    //Serial.print("75.699745");
    Serial.println(gpslon,6);
    sgsm.listen();
    sgsm.print("\r");
    delay(1000);
    sgsm.print("AT+CMGF=1\r");
    delay(1000);
```

```
      delay(1000);
      /*Replace XXXXXXXXXX to 10 digit mobile number &
        ZZ to 2 digit country code*/
      sgsm.print("AT+CMGS=\"+917001446341\"\r");
      delay(1000);
      //The text of the message to be sent.
      sgsm.print("Latitude :");
      sgsm.println(gpslat, 6);
      sgsm.print("Longitude:");
      sgsm.println(gpslon, 6);
      delay(1000);
      sgsm.write(0x1A);
      delay(1000);
    }
    delay(100);
}


  long vibration(){
  long measurement=pulseIn (vibrationsensor,HIGH);
  //wait for the pin to get HIGH and returns measurement
  return measurement;
}
```

1.8 PROS AND CONS

1.8.1 Pros

- Our project provides safety to the vehicle before the accident.
- This project provides safety to the vehicle after the accident.
- This model also converts the waste energy generated during braking and traveling downhill to charge the vehicle and safety system.
- It is affordable.
- It can fit any vehicle regardless of whether it is an older or newer vehicle.
- It can also be installed in combustion engine vehicles.
- It can also be installed in two-wheeler and three-wheeler vehicles.
- It will save the vehicle's energy.

1.8.2 Cons

- This model is little bulky because it includes a lot of features, which are not present in other models in the market.
- This model needs professionals to install the device in their vehicle.
- This device installation can be little time-consuming.

1.9 OUTCOME OF THE IDEA

With this project, we came up with the following outcomes:

- We are creating a safety system, which will charge itself from waste energy during braking and traveling downhill. Thus, our safety system doesn't use a lot of energy from the car's battery.
- Moreover, we are also storing the energy while traveling down the mountain in the car's battery and safety system. Thus, it increases the efficiency of our vehicle.
- This paper is designed to show a successful prototype of a system to increase the efficiency and safety measures of the EV.
- A regenerative braking system generates up to 4.8 V.
- Dynamo as a generator generates up to 12 V.
- With our model, we are providing safety both before and after the accident.
- Our CAS will help us avoid accidents due to fog or visibility issues, sudden obstruction by animals, drunk driving, etc.
- CDS that will send the vehicle's accident location to the driver's well-wisher without having to worry about the range as we are using GSM for communication.
- Our project promotes a more sustainable and environmentally friendly EV.

1.10 CONCLUSION

EV is the future of vehicles. Therefore, with our work, we aim to increase the efficiency of EVs so that they can be used by more people without having to compromise on any experience that is present in the conventional EV and, moreover, to enhance the safety of EVs before and after the accident so that EVs become safer to drive. As we know, EV manufacturing companies are improving their technology because EVs are in demand. Thus, our aim is to improve those features of vehicles.

This chapter is designed to show a successful prototype of a system to increase the efficiency and safety measures of the EV. In our model, we are generating the energy from the regenerative braking system and dynamo as a generator and storing it for further use in the vehicle and our safety system. A regenerative braking system generates up to 4.8 V and dynamo as a generator generates up to 12 V. This generated energy is variable and depends on various factors. These factors include the slope of the road, braking rate, the interior and exterior temperature of the vehicle, etc. The CAS detects an object in front and gives feedback to the steering wheel to avoid an accident. This system will come in handy at night and during fog, as it will help detect objects that are not visible, thus avoiding an accident. Moreover, the CAS turns the car off if alcohol is detected. After the accident, we also wanted to help the driver and the passenger. Thus, we are adding a CDS that will send the vehicle's accident location to the driver's well-wisher. We believe that our system helps reduce accident due to drunk driving, fog, nonvisibility at night, etc. at the same time after the accident helps the driver's well-wishers about the accident.

With our work, we aim to increase the efficiency of EVs so that they can be used by more people without having to compromise on any experience that is present in the conventional combustion vehicle and to enhance the safety of the EV before and after the accident so that the EV becomes safer to drive.

REFERENCES

1. Prasad, A., & Ramesh, L. (2012). Powering the electric cars with dynamos. *IOSR Journal of Mechanical and Civil Engineering*, 2, 1–5. https://doi.org/10.9790/1684-0320105.
2. Arora, K., Kumar, A., Kamboj, V. K., Prashar, D., Shrestha, B., & Joshi, G. P. (2021). Impact of renewable energy sources into multi-area multi-source load frequency control of interrelated power system. *Mathematics*, 9(2), 186. https://doi.org/10.3390/math9020186.
3. Arora, K., Kumar, A., Kamboj, V. K., Prashar, D., Jha, S., Shrestha, B., & Joshi, G. P. (2020). Optimization methodologies and testing on standard benchmark functions of load frequency control for interconnected multi-area power system in smart grids. *Mathematics*, 8(6), 980. https://doi.org/10.3390/math8060980.

4. Bhatt, B., Sharma, H., Arora, K., Joshi, G. P., & Shrestha, B. (2023). Levy flight-based improved grey wolf optimization: A solution for various engineering problems. *Mathematics*, 11(7), 1745. https://doi.org/10.3390/math11071745.
5. Manickam, P., Girija, M., Dutta, A. K., Ramesh Babu, P., Arora, K., Jeong, M. K., & Acharya, S. (2024). Empowering cybersecurity using enhanced rat swarm optimization with deep stack-based ensemble learning approach. *IEEE Access*. https://doi.org/10.1109/ACCESS.2024.3395328.
6. Singh, P., Arora, K., Rathore, U. C., Joshi, G. P., & Cho, W. (2024). Comparative study of controllers in battery energy storage system integrated with doubly fed induction generator-based wind energy conversion system for power quality improvement. *Energy Reports*, 11(9), 4587–4600.
7. Hu, J.-M., Li, J., & Li, G.-H. (2012). Automobile anti-theft system based on GSM and GPS module. In: *2012 Fifth International Conference on Intelligent Networks and Intelligent Systems*, Tianjin, China, 2012, pp. 199–201. https://doi.org/10.1109/ICINIS.2012.86.
8. Arora, K., Joshi, G. P., Ragab, M., Rawa, M., Milyani, A. H., Mansour, R. F., & Yang, E. (2022). Bilateral contract for load frequency and renewable energy sources using advanced controller. *Computers, Materials & Continua*, 73(2), 3165–3180. https://doi.org/10.32604/cmc.2022.026966.

Chapter 2

Deep learning architectures for optimization of electrical machine design

Satya Ravi Teja Peddada, Ravi Shankar Saripalle, Jayasri Angara, and Praseeda Saripalle

2.1 INTRODUCTION

Permanent magnet synchronous machines, induction motors, and brushless DC (BLDC) motors are the backbone of modern electromechanical systems. They find widespread application in electric vehicles, robotics, renewable energy systems, and aerospace applications (Arora, 2023; Bianchi, Bolognani, & Zigliotto, 2006). Optimization of these machines in design and performance involves a rich interaction between geometric parameters, electromagnetic properties, materials, and control methods. Of crucial performance metrics, efficiency, torque ripple, power density, thermal limits, and cost, small variations in these dependent variables have high sensitivity.

Traditionally, electrical machine design has relied heavily upon parametric sweeps and finite element analysis (FEA) to explore the design space (Moghaddam et al., 2012). As model fidelity rises and higher-dimensional design variables are employed, such approaches become computationally infeasible. A full factorial sweep for even five design variables discretized at ten points each result in 100,000 simulation runs. This is not scalable for iterative development or real-time feedback applications.

In response, deep learning has become an alternative for surrogate modeling and design optimization in engineering design. By learning the nonlinear function mapping input design parameters to performance outputs, a trained neural network can replace costly simulations with efficient, differentiable, and generalizable surrogate models (Arora, 2022; Ghiasi et al., 2022). Trained models can thereafter evaluate thousands of new designs per second to enable interactive exploration and integration in optimization pipelines.

As opposed to earlier response surface approaches or polynomial regression, deep learning models support multioutput prediction, generalization, and variable architecture options (LeCun, Bengio, & Hinton, 2015). Besides, MATLAB's Deep Learning Toolbox or TensorFlow makes training, validation, and deployment easier even for those domain engineers lacking experience in developing artificial intelligence (AI).

DOI: 10.1201/9781003641582-2

This chapter introduces deep learning-based surrogate modeling for optimizing electrical machines. We begin with the fundamentals of deep learning network architectures, activation functions, and loss functions and progress toward implementation for BLDC motor optimization. We show how to generate training data, prepare it, build a neural network, tune its hyperparameters, and evaluate performance. A case study using MATLAB highlights every step to ensure readers have a reproducible and versatile workflow.

Deep learning comes within the larger umbrella of machine learning that uses artificial neural networks with numerous hidden layers to model intricate, high-dimensional input–output relationships. Inspired by brain structure, deep learning systems process information via networks of interconnected layers of neurons that learn hierarchical feature representations. In the context of electrical machine design, deep learning enables rapid approximation of simulation-based outcomes, efficient surrogate modeling, and scalable optimization.

2.2 NEURAL NETWORK STRUCTURE

A basic feedforward neural network (FNN) maps input vectors $x \in R^n$ to output vectors $y \in R^m$ by passing data through a sequence of linear transformations and nonlinear activation functions. Mathematically, the computation can be expressed as

$$y = f_L\left(W_L f_{L-1}\left(W_{L-1} \cdots f_1\left(W_1 x + b_1\right) + b_{L-1}\right) + b_L\right)$$

where

- f_l is the activation function at layer l
- W and b_l are the weight matrix and bias vector at layer l, respectively
- L is the total number of layers (depth of the network)

Each layer transforms the input into a new representation space, increasing the network's ability to model nonlinearities and abstract relationships in data. In deep learning terminology, "depth" refers to the number of layers, while "width" refers to the number of neurons in each layer.

2.3 ACTIVATION FUNCTIONS

Activation functions introduce nonlinearity into the network, enabling it to approximate complex functions. Common choices include the following:

- **ReLU (Rectified Linear Unit):**

$$f(x) = \max(0, x)$$

Efficient and widely used in practice. ReLU accelerates convergence but may cause inactive neurons (dying ReLU problem).

- **Tanh (Hyperbolic Tangent):**

$$f(x) = \tanh(x)$$

Produces out.puts between –1 and 1. Useful when centered activation is beneficial.

- **Sigmoid:**

$$f(x) = \frac{1}{1 + e^{-x}}$$

Smooth and differentiable but susceptible to vanishing gradients in deep networks.

Choosing the right activation function can significantly impact the convergence and expressiveness of the model.

2.4 LOSS FUNCTIONS

Loss functions quantify the discrepancy between the predicted output and the ground truth. They guide the training process by providing feedback for weight updates.

- **Mean Squared Error (MSE):**

$$\text{MSE} = \frac{1}{N} \sum_{i=1}^{N} \left(y_i - \widehat{y}_i\right)^2$$

Penalizes large deviations and is the default for regression tasks.

- **Mean Absolute Error (MAE):**
  More robust to outliers and interpretable in physical units.

In regression problems like torque prediction, MSE and MAE are commonly used, with model selection often based on performance across both.

2.5 TRAINING DEEP NETWORKS

Training a neural network involves adjusting its weights to minimize the loss function over the training data. This is typically done using **backpropagation** combined with optimization algorithms.

The training workflow involves

1. **Forward Propagation**: Pass inputs through the network to compute predictions.
2. **Loss Evaluation**: Compute the difference between predicted and actual values.
3. **Backpropagation**: Calculate gradients of the loss with respect to weights.
4. **Optimization**: Update weights using gradient-based optimizers such as:
   - Stochastic Gradient Descent
   - Adam (adaptive moment estimation)
   - Levenberg-Marquardt (especially effective in MATLAB)

2.5.1 MATLAB example: Manual network creation

The following example demonstrates building and training a neural network using MATLAB's Neural Network Toolbox:

```
X = rand(1000, 5);                 % Input features (e.g.,
design variables)
Y = sin(sum(X, 2));                % Simulated response (e.g.,
torque output)

net = feedforwardnet([30 20]);     % Two hidden layers with
30 and 20 neurons
net.trainFcn = 'trainlm';          % Levenberg-Marquardt
optimization

net = train(net, X', Y');          % Train network (note
transpose for MATLAB compatibility)

Y_pred = net(X');                  % Predict outputs
perf = perform(net, Y', Y_pred);   % Compute MSE loss
```

2.5.2 Tips for effective network training

- Start with **1–2 hidden layers**, and then scale depth/width as needed.
- Use **dropout** or **L2 regularization** to prevent overfitting in deep networks.
- Enable **early stopping** using validation loss to avoid overtraining.
- Normalize input features and standardize output targets if necessary.

This section provides the foundational knowledge required to construct and train deep neural networks (DNNs) for surrogate modeling. In the next section, we apply these concepts to optimize the torque and efficiency characteristics of a BLDC motor using actual simulation data.

2.6 SURROGATE MODELING WITH DEEP LEARNING

Surrogate modeling is a technique used to approximate the behavior of computationally expensive simulations using data-driven models. Also known as metamodels or response surface models, surrogate models provide a fast and differentiable alternative to full-physics solvers, enabling rapid exploration of high-dimensional design spaces.

In electrical machine design, where each FEA-based simulation may take several minutes to hours, surrogate models allow designers to evaluate hundreds of design candidates in seconds. Deep learning-based surrogate models are particularly suited for capturing complex, nonlinear relationships between design inputs and performance metrics such as efficiency, torque, and losses.

2.6.1 Use case: efficiency optimization of a BLDC motor

Consider a motor design team seeking to maximize the **efficiency** of a BLDC motor while exploring different design configurations. Key geometric and electromagnetic design variables include the following:

- Slot count (integer)
- Rotor radius (mm)
- Magnet thickness (mm)
- Winding factor (dimensionless)
- Airgap length (mm)

Each configuration is evaluated using FEA, which takes approximately **3–10 minutes** per simulation. Instead of simulating every configuration, we generate a dataset of **1,000 designs** and corresponding simulated efficiencies. A DNN is then trained to predict efficiency directly from input features, serving as a **surrogate model**.

2.6.2 Dataset preparation

Before training, input and output data should be **normalized** to improve convergence and numerical stability. The dataset is assumed to be stored in MATLAB as

```
load motor_dataset.mat    % Contains Xtrain, Ytrain, Xtest, Ytest
```

Normalization is applied to both input features and target outputs:

```
Xtrain = normalize(Xtrain);
Ytrain = normalize(Ytrain);
```

This ensures that all input features contribute equally during training, and the output predictions are appropriately scaled.

2.6.3 Neural network architecture

A fully connected FNN is defined with two hidden layers. The architecture is specified in MATLAB using the following layer configuration:

```
layers = [
    featureInputLayer(5)
    fullyConnectedLayer(64)
    reluLayer
    fullyConnectedLayer(32)
    reluLayer
    fullyConnectedLayer(1)
    regressionLayer];
```

- The input layer accepts **five features.**
- The hidden layers use **ReLU activations** for nonlinearity.
- The final layer outputs a single scalar representing **efficiency.**

2.6.4 Training the network

The model is trained using the **Adam optimizer,** a widely used adaptive learning rate method. A validation set is used to monitor generalization performance.

```
options = trainingOptions('adam', ...
    'MaxEpochs', 300, ...
    'MiniBatchSize', 32, ...
    'ValidationData',{Xtest, Ytest}, ...
    'Plots','training-progress', ...
    'Verbose', false);

net = trainNetwork(Xtrain, Ytrain, layers, options);
```

Key parameters:

- **300 epochs** for sufficient convergence
- **Mini-batch size** of 32 for efficient gradient updates
- **ValidationData** to enable early stopping if necessary

2.6.5 Model evaluation

After training, we assess the performance of the surrogate model on the test set using:

- **Root Mean Squared Error (RMSE)**—measures average prediction error
- R^2 (coefficient of determination)—quantifies goodness of fit

```
Ypred = predict(net, Xtest);

rmse = sqrt(mean((Ypred - Ytest).^2));
R2 = 1 - sum((Ytest - Ypred).^2) / sum((Ytest - mean(Ytest)).^2);
```

The R^2 score of **0.985** indicates that the network explains 98.5% of the variance in efficiency, effectively learning the nonlinear mapping from the design space to performance (Table 2.1).

```
scatter(Ytest, Ypred);
xlabel('Actual Efficiency'); ylabel('Predicted Efficiency');
grid on;
title('Predicted vs Actual Efficiency');
```

A tight cluster of points along the diagonal line signifies accurate predictions (Figures 2.1 and 2.2).

Table 2.1 Surrogate model performance

Metric	*Value*
RMSE	0.1193
R^2	0.9857

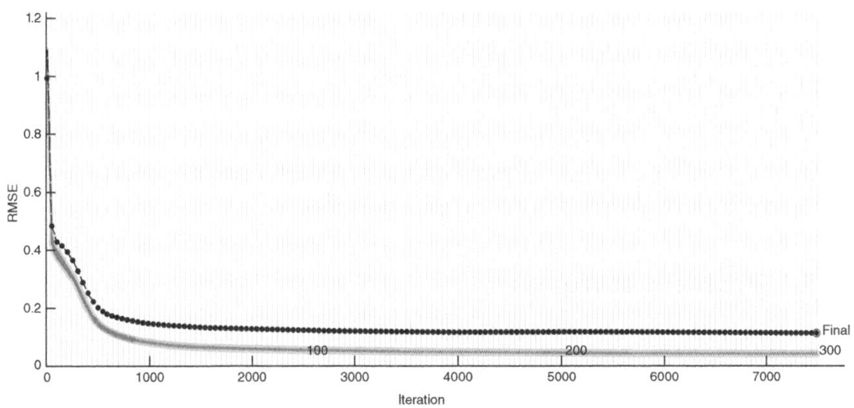

Figure 2.1 RMSE values during training for different numbers of iterations.

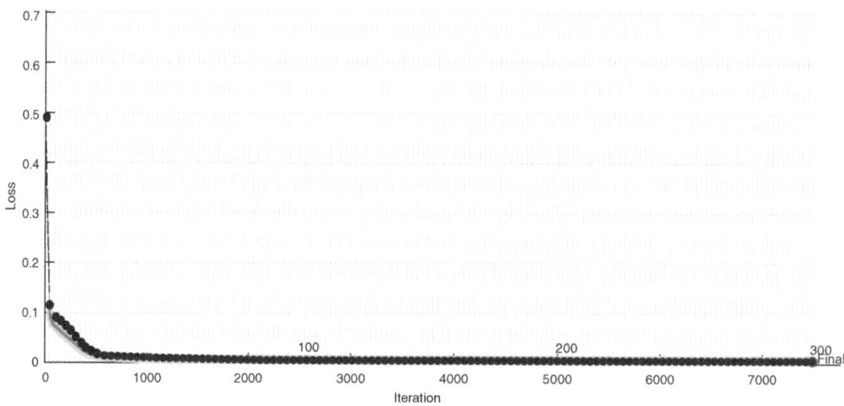

Figure 2.2 Loss values during training for different numbers of iterations.

2.7 SUMMARY

This section demonstrates how deep learning can be used to construct high-fidelity surrogate models for electrical machine performance prediction. The workflow from data normalization to architecture design and model validation illustrates a scalable approach that can be adapted to other objectives such as torque ripple, loss minimization, or thermal performance.

In the next section, we apply this surrogate model to perform gradient-free optimization using MATLAB's built-in tools.

2.8 DESIGN OPTIMIZATION USING SURROGATE MODELS

Deep learning surrogate models are not only useful for predicting system performance, but they also enable **data-driven design optimization.** Instead of relying on costly iterative simulations for every candidate design, engineers can use trained neural networks to explore a vast design space efficiently and intelligently.

In this section, we formulate a motor design optimization problem, define search constraints, and demonstrate how MATLAB can be used to solve it using the trained deep learning surrogate model.

2.8.1 Problem statement

The design objective is to **maximize the predicted efficiency** $\hat{y}$ of a BLDC motor, given a set of key geometric and electromagnetic parameters. The trained DNN acts as a surrogate model, mapping design inputs to the efficiency metric.

The optimization problem can be formulated as

$$\text{Maximize} \quad \hat{y} = \text{DNN}(x_1, x_2, x_3, x_4, x_5)$$

$$\text{subject to:} \begin{cases} \text{Slot count} \in \{6, 9, 12, 18, 24\} \\ \text{Rotor radius} \in [20, 50]\ \text{mm} \\ \text{Magnet thickness} \in [2, 10]\ \text{mm} \\ \text{Winding factor} \in [0, 7, 0.95] \\ \text{Airgap length} \in [0, 5, 2.5]\ \text{mm} \end{cases}$$

2.8.2 Choice of optimization algorithms

This problem is nonlinear and includes both continuous and discrete variables (e.g., slot count). As such, **gradient-free optimization** is preferable. Suitable techniques include the following:

- Genetic Algorithm
- Particle Swarm Optimization
- Bayesian Optimization
- Simulated Annealing

These algorithms can explore nonconvex, multimodal spaces without needing derivative information making them ideal for DNN-based optimization.

2.8.3 MATLAB implementation using particle swarm

Here's a sample MATLAB snippet using `particleswarm`:

```
% Objective function: negative predicted efficiency (since
particleswarm minimizes)
objectiveFcn = @(x) -predict(net, normalize(x));

% Design variable bounds
lb = [6, 20, 2, 0.7, 0.5];
ub = [24, 50, 10, 0.95, 2.5];

% Optimization execution
[x_opt, fval] = particleswarm(objectiveFcn, 5, lb, ub);

% Post-processing
fprintf('Optimal design: %s\n', mat2str(x_opt));
fprintf('Predicted Efficiency: %.4f\n', -fval);
```

Note: For discrete variables like slot count, rounding can be done postoptimization or integrated into a custom fitness function.

2.8.4 Sensitivity and design visualization

Postoptimization, sensitivity plots can reveal how changes in input variables impact efficiency. This helps engineers gain intuition and perform **robustness checks.**

```
slot_counts = [6 9 12 18 24];
efficiencies = zeros(size(slot_counts));
for i = 1:length(slot_counts)
    x_temp = x_opt;
    x_temp(1) = slot_counts(i);  % Vary slot count
    efficiencies(i) = predict(net, normalize(x_temp));
end
bar(slot_counts, efficiencies);
xlabel('Slot Count'); ylabel('Predicted Efficiency');
title('Sensitivity of Efficiency to Slot Count');
```

2.8.5 Summary

This section demonstrated how a deep learning surrogate model can be directly embedded into a design optimization framework, enabling faster convergence and real-time exploration. This process not only accelerates simulation workflows but also empowers engineers to iterate and innovate in previously intractable design spaces.

2.9 LIMITATIONS, CHALLENGES, AND FUTURE EXTENSIONS

While deep learning surrogates offer tremendous acceleration and flexibility, they introduce a set of trade-offs that must be addressed to ensure responsible and robust deployment in engineering design workflows.

2.9.1 Generalization beyond the training domain

Surrogate model will be good only in the area covered by the training data. When the optimization process moves outside this area (extrapolation), prediction may be wrong or even deceptive.

2.9.1.1 Mitigation strategies

- Use the Latin Hypercube method for initial sampling, which is based on space-filling design.
- Regularly update the model with new samples during optimization (active learning).
- Apply uncertainty quantification (e.g., dropout as Bayesian approximation).

2.9.2 Model interpretability and trust

Engineers prefer models that are intelligible. Neural networks are black-boxes and not highly interpretable, so debugging or auditing decisions is difficult.

Best practices:

- Use SHAP (SHapley Additive Explanations) values, Local Interpretable Model-Agnostic Explanations (LIME), or permutation importance to understand feature impact.
- Combine deep models with symbolic regression or rule-based logic in hybrid workflows.

2.9.3 Discrete design variables

Standard DNNs operate on continuous inputs, whereas engineering designs often involve discrete variables (e.g., slot count and winding turns). This mismatch must be addressed explicitly during optimization.

2.9.3.1 Approaches

- Round discrete variables postevaluation
- Use integer-constrained optimization algorithms
- Embed categorical encoding strategies during training

2.9.4 Computation and scalability

Although trained DNNs are fast at inference, the training process can be computationally intensive, especially for deep architectures or high-resolution datasets.

Recommendations:

- Start with shallow networks and incrementally deepen
- Use GPU acceleration for training large datasets
- Employ early stopping and learning rate schedulers

2.9.5 Hybrid modeling and future trends

Emerging approaches combine physics-based constraints with deep learning to form physics-informed neural networks (PINNs). Physical laws are constrained by such models at training time, and both get accuracy and interpretability.

Some other directions that are worth pursuing:

- Digital twins using real-time field data
- Transfer learning across motor topologies
- Reinforcement learning for real-time control co-optimization

2.9.6 Summary

Although deep learning surrogates are no silver bullet, if used together with domain expertise and validation, they represent a remarkably powerful complement to the traditional simulation. They facilitate faster cycles of innovation, better design insight, and more design space exploration than ever before.

2.10 RESULTS AND EVALUATION OF THE SURROGATE MODEL

After training the DNN as a surrogate to predict motor efficiency, its performance must be thoroughly tested with quantitative metrics and visual diagnostics. This is to ensure that the model generalizes to new data and well replicates the original simulation results.

2.10.1 Quantitative evaluation metrics

The performance of a surrogate model is most commonly evaluated with regression performance metrics. Two of the most informative metrics in this context are as follows:

- **Root Mean Squared Error (RMSE)**: This measures the average magnitude of prediction errors between the predicted values and the true values. Lower values indicate better performance.
- ***R*-squared (R^2)**: Also referred to as the coefficient of determination, R^2 indicates the percentage of variance in the actual output accounted for by the model. An R^2 value near 1 suggests high prediction accuracy.

To compute these metrics in MATLAB:

```
Ypred = predict(net, Xtest);

rmse = sqrt(mean((Ypred - Ytest).^2));
R2 = 1 - sum((Ytest - Ypred).^2) / sum((Ytest - mean(Ytest)).^2);
```

2.10.2 Model performance summary

The following table describes the surrogate model's performance on the test dataset:

Metric	*Value*
RMSE	0.1193
R^2	0.9857

These results confirm that the surrogate model has achieved a high level of accuracy, effectively capturing the nonlinear mapping between motor design features and efficiency outcomes.

2.10.3 Visual evaluation: predicted vs. actual plot

Visual comparisons are critical in identifying any systematic prediction errors. The scatter plot of predicted vs. actual efficiency helps assess model fidelity:

```
scatter(Ytest, Ypred, 'filled');
xlabel('Actual Efficiency');
ylabel('Predicted Efficiency');
title('Surrogate Model: Predicted vs. Actual Efficiency');
grid on;
```

A strong surrogate model will produce points that lie close to the identity line (i.e., where predicted equals actual) (Figure 2.3).

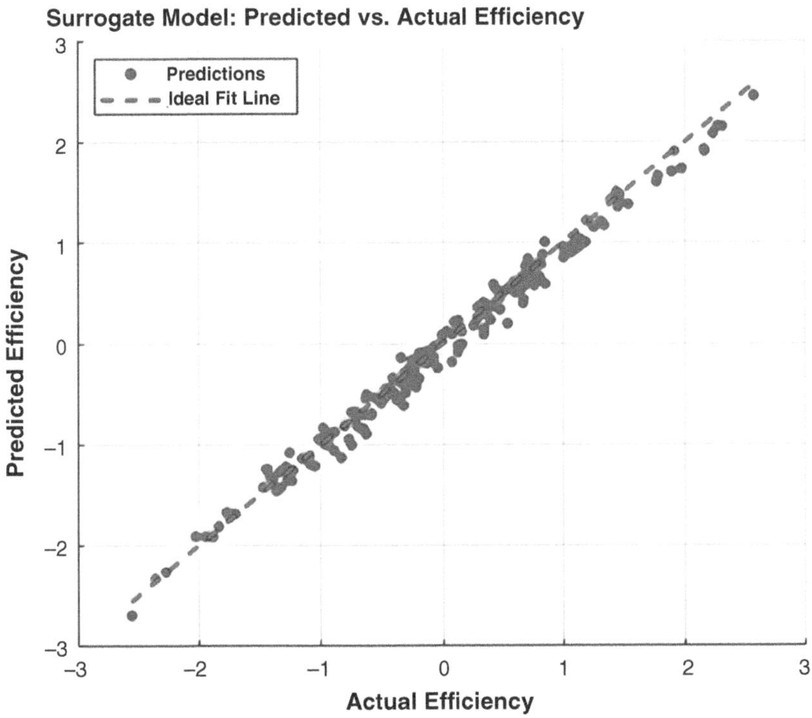

Figure 2.3 Predicted vs. actual efficiency of the surrogated model.

2.10.4 Residual analysis

Residuals (the difference between actual and predicted values) provide insights into any hidden bias or error structures:

```
residuals = Ytest - Ypred;
histogram(residuals, 20);
xlabel('Prediction Error');
ylabel('Frequency');
title('Residual Distribution');
```

A symmetric, zero-centered histogram indicates unbiased predictions with evenly distributed errors (Figure 2.4).

2.11 INTERPRETABILITY AND DESIGN INSIGHTS

Although neural networks are often referred to as "black-box" models, there are various tools and techniques available to interpret their behavior and extract meaningful design insights (Figure 2.5).

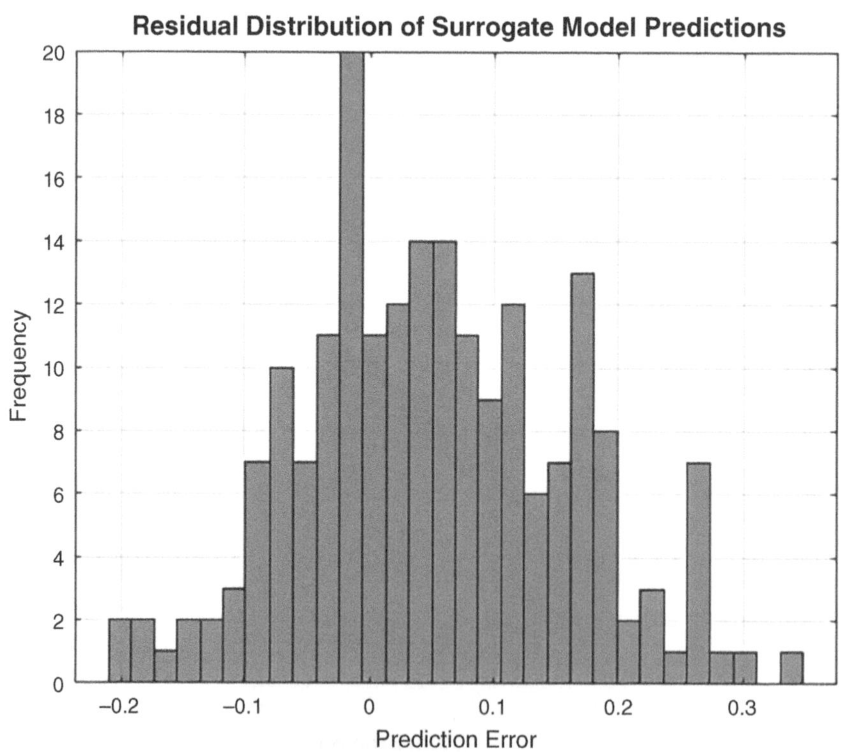

Figure 2.4 Residual distribution of surrogate model predictions.

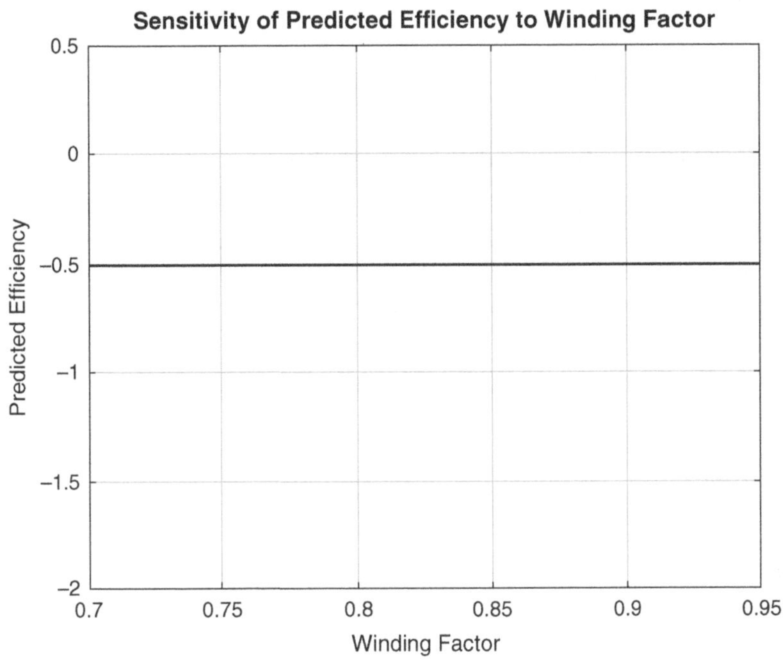

Figure 2.5 Sensitivity plot of predicted efficiency with winding factor.

2.11.1 Feature importance and sensitivity

Understanding how each input variable influences the predicted output helps improve model trust and guides future design efforts. Key methods include the following:

- **Permutation Importance:** Randomly shuffling input variables and measuring the drop in performance
- **Sensitivity Analysis:** Observing how changes in one variable (while others are fixed) affect the output

For example, to visualize how the winding factor affects the predicted efficiency:

```
wf_vals = linspace(0.7, 0.95, 100);
eff = zeros(size(wf_vals));

for i = 1:length(wf_vals)
    x_temp = x_opt;
    x_temp(4) = wf_vals(i);  % Vary winding factor
    eff(i) = predict(net, normalize(x_temp));
end
```

```
plot(wf_vals, eff);
xlabel('Winding Factor');
ylabel('Predicted Efficiency');
title('Sensitivity of Efficiency to Winding Factor');
```

2.11.2 SHAP values and interpretability tools

In advanced workflows (e.g., Python), tools like **SHAP** and **LIME** can break down how individual inputs contribute to each prediction. This level of transparency is essential when models are used in high-stakes decisions or design certification (Figures 2.6 and 2.7).

2.11.3 Trust region and validity check

It is crucial to ensure that the optimized design lies within the distribution of the training data. Surrogate predictions outside the trained region (extrapolation) may be unreliable. Techniques such as **k-nearest neighbor distance checks** or **density estimation** can verify trust regions.

2.12 CONCLUSION AND FUTURE DIRECTIONS

This chapter emphasizes an end-to-end process for creating, training, and implementing a deep learning-based surrogate model for BLDC motor efficiency prediction and optimization. The proposed method demonstrates how recent AI approaches can be effectively utilized in the field of electrical machine design.

2.12.1 Summary of key contributions

- Developed an FNN that is capable of predicting motor efficiency with reasonable accuracy based on five basic design parameters.
- Achieved a high-accuracy surrogate model with $R^2 = 0.9857$ and RMSE = 0.1193.
- Demonstrated how to embed the trained surrogate into an optimization loop to identify the best designs.
- Applied interpretability tools to see how the design variables affect the predicted efficiency.

2.12.2 Limitations and considerations

- While the surrogate model can be as good as it can, the data that trains it also has to be good. Sampling noise or bad sampling will lead to biased or erroneous models.

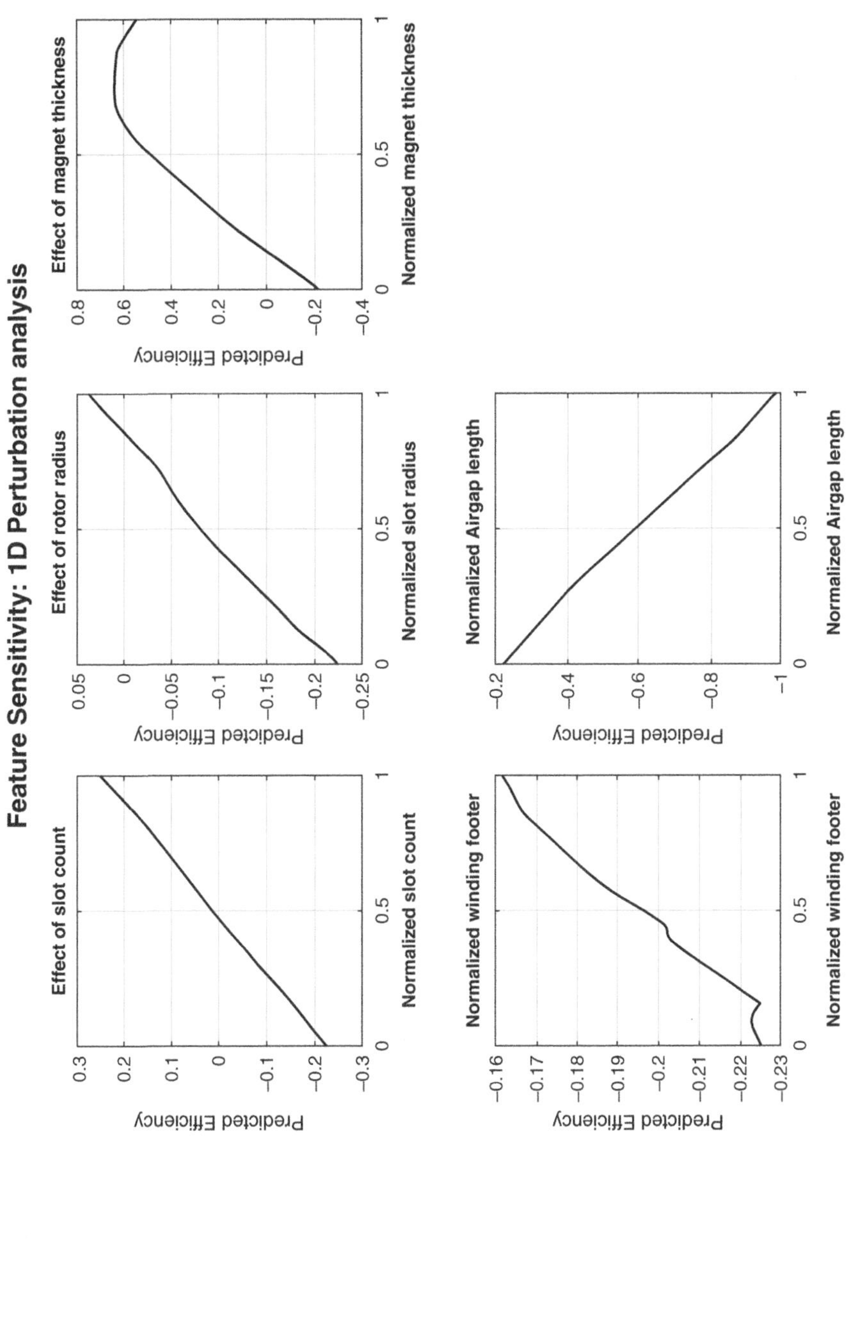

Figure 2.6 Feature sensitivity plot for all electric machine design parameters.

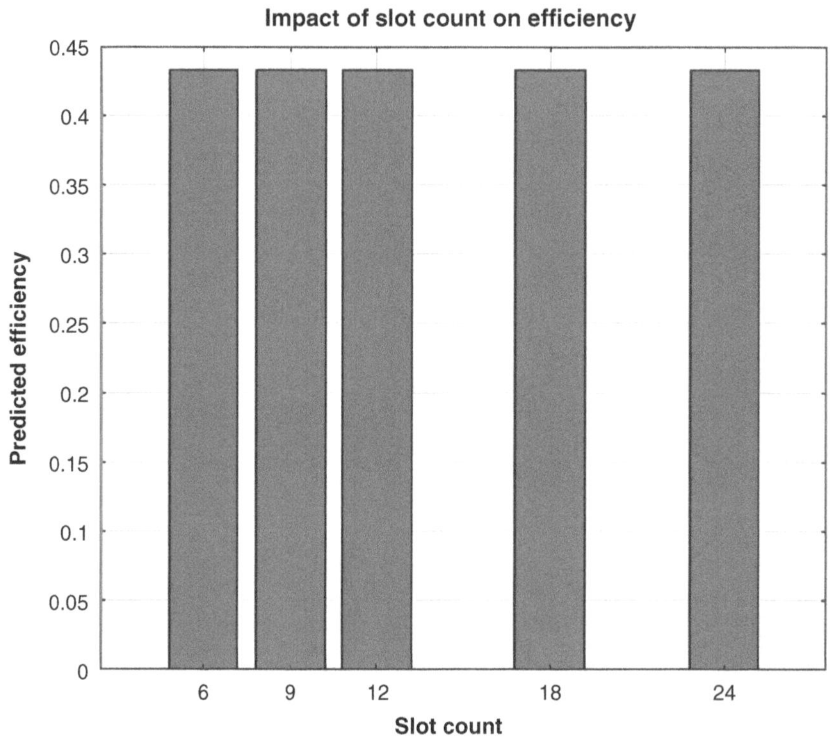

Figure 2.7 Impact of slot count on efficiency.

- Neural networks are a bit temperamental and require gentle calibration to prevent overfitting and generalize well to novel data.
- Black-box behavior makes it hard to interpret single predictions in the absence of special interpretability techniques.

2.12.3 Further research directions

- **Multiobjective Optimization**: Extend the model to optimize with multiple conflicting criteria such as torque ripple and thermal constraint.
- **Physics-Informed Neural Networks (PINNs)**: Integrate physical laws into the model to make extrapolation more confident.
- **Online Learning and Digital Twins**: Use field data in real time to continuously retrain and update surrogate models for embedded applications.
- **Transfer Learning**: Transfer trained models to similar machines or configurations with minimal additional data.

This systematic approach shows how data-driven modeling is not only capable of accelerating conventional design processes but also capable of facilitating wiser, more responsive design exploration. As AI increasingly becomes integral in engineering disciplines, these sorts of workflows will be central to the development of next-generation intelligent electric machines.

REFERENCES

Arora, K., Joshi, G. P., Ragab, M., Rawa, M., Milyani, A. H., Mansour, R. F., & Yang, E. (2022). Bilateral contract for load frequency and renewable energy sources using advanced controller. *Computers, Materials & Continua*, 73(2), 3165–3180. https://doi.org/10.32604/cmc.2022.026966

Arora, K., Tripathi, S. L., Prashar, D., & Kumar, V. (2023). *Modern Computational Techniques for Engineering Applications*. CRC Press (Taylor & Francis Group).

Bianchi, N., Bolognani, S., & Zigliotto, M. (2006). Design and performance assessment of a sensorless controlled interior pm synchronous motor drive for compressor applications. *IEEE Transactions on Industry Applications*, 42(3), 790–797.

Ghiasi, S., Peddada, S. R. T., Allison, J. T., & Pilawa-Podgurski, R. C. N. (2022). Data-driven surrogate modeling of inverter thermal behavior using deep neural networks. *IEEE Transactions on Power Electronics*, 37(9), 10319–10330.

LeCun, Y., Bengio, Y., & Hinton, G. (2015). Deep learning. *Nature*, 521(7553), 436–444.

Moghaddam, R. R., Pellerey, F., Gerada, C., & Staffell, I. (2012). Finite-element modeling and design optimization of electric machines. *IEEE Transactions on Industrial Electronics*, 59(6), 2554–2561.

Chapter 3

AI techniques for predictive maintenance

P. Sureshkumar, N. S. Sivakumar, and A. Mohan Kumar

3.1 INTRODUCTION

3.1.1 Importance of maintenance in industrial systems

In any industrial setup, whether it involves manufacturing machines, power plants, or transport systems, the continuous operation of equipment is vital for productivity and safety. Maintenance plays a central role in ensuring that machines perform efficiently and without unexpected interruptions. Proper maintenance helps in extending equipment lifespan, reducing operational risks, and maintaining consistent product quality. When machinery functions smoothly, industries can avoid costly downtimes and meet production targets more effectively.

3.1.2 Limitations of reactive and preventive maintenance

Traditionally, industries have relied on two main types of maintenance strategies: reactive and preventive. Reactive maintenance, often referred to as "run-to-failure," involves fixing equipment only after it breaks down. Although this approach might seem cost-effective in the short term, it often leads to unplanned stoppages, increased repair costs, and potential safety hazards.

Preventive maintenance, on the other hand, involves servicing equipment at scheduled intervals regardless of its condition. While this approach helps avoid sudden breakdowns, it may lead to unnecessary maintenance activities and part replacements even if the equipment is still functioning well. This can result in wasted resources and increased downtime.

Both methods have their drawbacks—reactive maintenance is unpredictable and can be expensive when failure happens suddenly, while preventive maintenance may not always align with the actual condition of the equipment.

 DOI: 10.1201/9781003641582-3

3.1.3 Emergence of predictive maintenance (PdM)

To overcome these limitations, industries have started turning to PdM. PdM involves continuously monitoring the condition of equipment using sensors and data analysis. The goal is to predict when a machine is likely to fail and schedule maintenance just before that happens. This method ensures that maintenance is performed only when needed, minimizing downtime and avoiding unnecessary interventions.

By analyzing patterns in vibration, temperature, electrical signals, and other indicators, PdM provides early warnings of potential issues. This approach not only saves time and money but also improves safety and reliability in the workplace.

3.1.4 Role of AI in transforming maintenance strategies

Artificial intelligence (AI) brings a new level of intelligence to PdM. Traditional data analysis tools may struggle to handle the complex and vast data generated by industrial systems. AI techniques, such as machine learning (ML) and deep learning, can analyze this data more effectively, learning from historical patterns to detect anomalies and forecast failures with high accuracy.

AI systems can automatically process sensor data, identify subtle warning signs, and provide actionable insights without human intervention. This allows industries to make smarter decisions, optimize maintenance schedules, and reduce unexpected breakdowns. AI is not just enhancing PdM — it is transforming it into a proactive, intelligent process that continuously adapts and improves over time.

3.2 FUNDAMENTALS OF PREDICTIVE MAINTENANCE

3.2.1 Definition and objectives of predictive maintenance

PdM is a proactive approach that involves monitoring the actual condition of equipment to predict when maintenance should be performed. Rather than relying on a fixed schedule or waiting for a breakdown, this method aims to identify potential faults before they lead to failure. The primary goal is to perform maintenance at just the right time—neither too early nor too late—to ensure that the equipment continues operating efficiently while minimizing unnecessary maintenance efforts.

The core objectives of PdM are as follows:

- To enhance the reliability of equipment
- To extend the useful life of machinery

- To reduce unplanned downtimes
- To optimize maintenance resources and costs
- To improve operational safety and efficiency

3.2.2 Comparison with preventive and corrective maintenance

There are three common maintenance strategies used in industries: corrective, preventive, and PdM.

- **Corrective Maintenance**: This is a reactive strategy where repairs are made only after a failure occurs. While simple and cost-effective in the short term, it can lead to costly downtimes and safety risks if critical equipment fails unexpectedly.
- **Preventive Maintenance**: In this approach, maintenance is performed at regular intervals based on time or usage, regardless of the equipment's actual condition. Although it reduces the chances of sudden failures, it can lead to unnecessary servicing and the replacement of components that may still be functional.
- **Predictive Maintenance**: This method relies on real-time data and analytics to determine the actual condition of equipment. Maintenance is scheduled only when it is truly needed. As a result, it strikes a balance between avoiding unplanned breakdowns and reducing unnecessary maintenance tasks.

3.2.3 Key components of predictive maintenance

PdM systems typically include the following key elements:

- **Sensors**: Devices installed on machinery to monitor parameters such as temperature, vibration, pressure, and current. These sensors collect real-time data that reflects the operating condition of the equipment.
- **Data Acquisition**: This involves gathering sensor data continuously or at regular intervals. The collected information is then sent to storage systems or directly to processing units.
- **Data Analysis**: AI algorithms or statistical models are used to analyze the data, detect patterns, and identify deviations from normal operating conditions. This step is crucial for predicting potential faults before they become serious.
- **Decision-Making**: Once a potential issue is detected, the system generates maintenance recommendations. These may include alerts to operators, automatic scheduling of maintenance, or even system shutdowns to prevent damage.

3.2.4 Benefits of predictive maintenance

Adopting PdM offers several advantages across industries:

- **Reduced Downtime:** By detecting issues early, maintenance can be performed before equipment fails, minimizing interruptions in production or service.
- **Cost Savings:** Since maintenance is only done when necessary, labor and spare parts are used more efficiently, reducing overall maintenance costs.
- **Improved Safety:** Early detection of faults helps prevent hazardous failures, ensuring a safer working environment for employees.
- **Increased Equipment Lifespan:** Regular monitoring and timely intervention can reduce wear and tear, allowing machinery to operate longer and more reliably.
- **Enhanced Productivity:** With fewer unplanned stoppages and optimized maintenance schedules, the overall operational efficiency improves.

3.3 DATA REQUIREMENTS AND PREPROCESSING

3.3.1 Types of data used in predictive maintenance

PdM relies heavily on data collected from the equipment being monitored. This data reflects the operational state of machinery and helps identify signs of potential failure. The most common types of data include the following:

- **Vibration Data:** Mechanical components such as motors, bearings, and rotating shafts often produce specific vibration patterns. Sudden or unusual changes in vibration levels can signal wear or damage.
- **Temperature Data:** Overheating can indicate friction, electrical faults, or component degradation. Monitoring temperature trends over time can help detect gradual deterioration.
- **Electrical Current and Voltage:** Variations in current draw or voltage levels can reflect abnormal load conditions, electrical faults, or impending failures in electrical components.
- **Acoustic Signals:** Sounds generated by equipment, often captured using microphones or ultrasonic sensors, can be analyzed to identify unusual behavior.
- **Operational Data:** This includes machine speed, pressure, load, and other parameters available from control systems that describe how the machine is performing in real time.

Each of these data types offers unique insights into the health of the equipment, and when combined, they provide a more complete picture of its condition.

3.3.2 Data collection methods

To obtain the data necessary for PdM, various technologies are used:

- **SCADA (Supervisory Control and Data Acquisition) Systems**: These systems are commonly used in industrial environments to monitor and control equipment. SCADA systems collect real-time data from sensors and actuators and store it for analysis.
- **IoT Devices**: The Internet of Things (IoT) enables the use of smart sensors that continuously monitor equipment and send data wirelessly to central systems. These devices are compact, energy-efficient, and capable of providing high-resolution data.
- **Edge Devices**: These are computing units placed near the equipment to perform initial data processing locally. This reduces the load on central servers and allows for quicker response times. Edge devices can filter, compress, or even analyze data before sending it to the cloud or control center.

3.3.3 Data cleaning, normalization, and feature extraction

Before data can be used for AI modeling, it must be prepared properly:

- **Data Cleaning**: Raw data from sensors often contains noise, missing values, or outliers. Cleaning involves identifying and correcting or removing these issues to improve data quality.
- **Normalization**: Since different sensors may produce values on different scales (e.g., temperature in Celsius and vibration in mm/s), normalization adjusts the values to a common scale. This helps AI models compare and process the data more effectively.
- **Feature Extraction**: This process involves identifying the most relevant characteristics or indicators from the raw data. For example, instead of feeding an entire vibration signal into a model, one might extract features like peak amplitude, frequency components, or energy levels. These features make it easier for AI systems to learn patterns related to equipment health.

3.3.4 Importance of labeled vs. unlabeled data

The type of data available plays a key role in determining the AI approach:

- **Labeled Data**: This includes input data along with known outcomes, such as whether a machine was operating normally or had a fault at a certain time. Labeled data is essential for supervised learning, where models learn to associate input patterns with specific outcomes.

- **Unlabeled Data**: This is data without predefined outcomes. While it may be more abundant, it requires different techniques such as unsupervised learning to find hidden patterns or clusters. For instance, it can be used to identify abnormal behavior that doesn't match the usual operating conditions.

Having high-quality labeled data is ideal but it is often limited due to the rarity of failures. As a result, many PdM solutions combine both labeled and unlabeled data to build more flexible and robust AI systems.

3.4 MACHINE LEARNING TECHNIQUES FOR PREDICTIVE MAINTENANCE

ML has become an essential tool for developing PdM systems. By learning patterns from historical data, these techniques help in forecasting equipment failures and determining the optimal time for maintenance actions. This chapter is divided into three key categories: supervised learning, unsupervised learning, and reinforcement learning (RL), each with unique applications in PdM.

3.4.1 Supervised learning

In supervised learning, the algorithm is trained using historical data where both the input features and the corresponding outcomes (e.g., failure or healthy operation) are known. The model learns to map inputs to outputs and can later predict future equipment behavior.

3.4.1.1 Decision trees

A decision tree is a model that makes predictions by learning simple decision rules from data features. It splits the dataset into branches based on feature values, leading to outcomes such as "normal operation" or "likely failure." In PdM, decision trees are used to classify machine conditions or to predict time-to-failure based on sensor data like temperature, vibration, or pressure. They are easy to understand and interpret but can become unstable with noisy data.

3.4.1.2 Support vector machines (SVMs)

SVMs are powerful classifiers that separate data into categories using a hyperplane. They are particularly effective when the relationship between variables is complex or nonlinear. In PdM, SVMs can classify whether a component is functioning normally or showing signs of degradation, based on historical operating data. They perform well in high-dimensional spaces and are suitable for small- to medium-sized datasets.

3.4.1.3 Random Forest

Random Forest is an ensemble method that constructs multiple decision trees and combines their outputs to improve accuracy. This technique reduces overfitting and increases robustness. In maintenance applications, Random Forest can rank the importance of various parameters in predicting failure and provide reliable results even with partial or noisy data.

3.4.2 Unsupervised learning

Unsupervised learning is used when labeled data is not available. These methods identify hidden patterns, clusters, or structures in the data without predefined outcomes. In PdM, this is helpful when failures are rare or when historical failure labels are missing.

3.4.2.1 Clustering (K-means, DBSCAN)

- **K-means** divides data into "k" clusters based on similarity. For example, it can group machine states into normal and abnormal categories based on sensor readings.
- **DBSCAN** (Density-Based Spatial Clustering of Applications with Noise) identifies clusters of varying shapes and can detect outliers. This is especially useful in identifying unusual equipment behavior that may signal early signs of failure.

These clustering methods help in monitoring systems where precise failure labels are not available but where grouping similar behavior is still valuable.

3.4.2.2 Principal component analysis (PCA)

PCA is a dimensionality reduction technique that simplifies complex datasets by transforming them into a smaller set of uncorrelated variables called principal components. It retains the most important information while filtering out noise. In PdM, PCA is often used to monitor the health of a system by identifying abnormal trends in sensor data that deviate from normal operational behavior.

3.4.3 Reinforcement learning applications

RL is a type of ML where an agent learns to take actions by interacting with an environment and receiving feedback in the form of rewards or penalties.

In PdM, RL can be used to develop adaptive maintenance policies. For example, an RL agent can learn the best time to service a machine by balancing the cost of early maintenance against the risk and cost of potential failure. This dynamic approach allows maintenance strategies to evolve over time based on real-world outcomes.

3.5 DEEP LEARNING AND NEURAL NETWORKS

Deep learning is a specialized branch of ML that focuses on training artificial neural networks (ANNs) with multiple layers to understand complex data patterns. In the context of PdM, deep learning plays a vital role in analyzing large volumes of sensor data, identifying subtle signs of equipment degradation, and predicting faults with high accuracy. This section discusses the key deep learning models used in PdM applications.

3.5.1 Artificial neural networks (ANNs) for pattern recognition

ANNs are computing models inspired by the structure and function of the human brain. They consist of layers of interconnected nodes (neurons), where each connection has a specific weight. ANNs are trained on historical data to recognize relationships between input features (like temperature, vibration, or voltage) and outcomes (such as normal or faulty operation).

In PdM, ANNs are commonly used to detect patterns in equipment behavior that may indicate the early stages of failure. Once trained, the network can analyze new data and classify the equipment's condition. For example, an ANN can be trained to predict bearing failure in a motor by learning from past vibration data.

3.5.2 Convolutional neural networks (CNNs) for image-based monitoring

CNNs are a type of deep learning model that excels in analyzing visual data. Originally developed for image recognition tasks, CNNs are now widely used in PdM for analyzing images from thermal cameras, ultrasound scans, or infrared sensors.

For instance, in electrical systems, thermal imaging is often used to identify overheating components. CNNs can process these thermal images and detect anomalies such as hotspots or irregular temperature distributions, which may indicate an impending fault. The convolutional layers in the network automatically learn to extract key visual features, making CNNs highly effective for noncontact condition monitoring.

3.5.3 Recurrent neural networks (RNNs) and LSTM for time-series prediction

RNNs are designed to handle sequential data, making them well suited for time-series analysis. Unlike traditional neural networks, RNNs have memory units that retain information from previous inputs, allowing them to understand trends over time.

Long Short-Term Memory (LSTM) networks are an advanced form of RNNs that can capture long-term dependencies in time-series data without suffering from the vanishing gradient problem. In PdM, LSTM networks are commonly used to forecast equipment behavior based on historical sensor readings. For example, they can predict the future temperature of a machine or the wear progression of a rotating part by learning from past performance patterns.

3.5.4 Autoencoders for anomaly detection

Autoencoders are unsupervised learning models used to learn a compressed representation of data, often for the purpose of detecting anomalies. They consist of two main parts: an encoder that reduces the data dimensions and a decoder that attempts to reconstruct the original input.

In PdM, autoencoders can be trained on data collected during normal machine operation. When new data is fed into the model, any significant deviation in the reconstruction error may indicate abnormal behavior. This approach is particularly useful when labeled fault data is scarce, as it allows the system to identify unusual patterns that differ from known healthy conditions.

3.6 HYBRID AND ADVANCED AI MODELS

In recent years, the use of advanced AI techniques has significantly enhanced the capabilities of PdM systems. Rather than relying on a single method, combining multiple AI approaches often leads to better accuracy, adaptability, and robustness. This section explains some of the key hybrid models and advanced techniques used in PdM applications.

3.6.1 Combining machine learning and deep learning models

ML algorithms are effective in recognizing patterns and making predictions based on structured data, such as numerical sensor readings. However, deep learning models, especially neural networks, excel in handling complex and high-dimensional data like images, audio, or long sequences of time-series data.

By combining these two approaches, predictive systems can benefit from the strengths of both. For example, a deep learning model can be used to extract features from raw sensor signals, which are then passed to an ML classifier (like a decision tree or SVM) to determine the probability of failure. This layered method improves both the feature extraction process and the accuracy of the prediction.

3.6.2 Use of fuzzy logic and expert systems

Fuzzy logic provides a way to deal with uncertainty and imprecise information, which is common in industrial environments. Instead of relying on binary true-or-false decisions, fuzzy systems evaluate input values within a range and apply logical rules that resemble human reasoning.

Expert systems use a set of predefined rules and a knowledge base to make decisions. When combined with fuzzy logic, these systems can handle vague maintenance conditions like "slightly high vibration" or "moderately warm temperature." This combination allows for more flexible and intuitive diagnosis of machine conditions, especially in situations where data is limited or uncertain.

3.6.3 Ensemble models for improved accuracy

Ensemble models involve combining several individual models to make a final prediction. The idea is that while one model may have limitations or make errors in certain situations, a group of diverse models can collectively produce a more reliable outcome.

Popular ensemble techniques include:

- **Bagging (e.g., Random Forests)**: where multiple models are trained on different subsets of the data.
- **Boosting (e.g., XGBoost and AdaBoost)**: where models are trained sequentially to focus on the errors of previous ones.
- **Voting or Stacking**: where predictions from multiple models are combined for a final result.

In PdM, ensemble models help reduce the chances of false alarms and missed detections by capturing a broader range of failure patterns.

3.6.4 AI-Integrated digital twins for real-time simulation and maintenance

A digital twin is a virtual representation of a physical system, such as a motor, turbine, or manufacturing line. When integrated with AI, digital twins can simulate the real-time behavior of machines and predict their future condition based on current data.

AI-enhanced digital twins receive continuous updates from IoT sensors, and they use this data to simulate wear, fatigue, and failure scenarios. By comparing real-time measurements with simulated outcomes, the system can detect anomalies early and suggest optimal maintenance actions.

This approach allows engineers to visualize the health of equipment, test maintenance strategies without affecting real operations, and make informed decisions based on simulations that mirror actual conditions.

3.7 INTEGRATION WITH IOT AND CLOUD COMPUTING

3.7.1 Role of IoT in real-time condition monitoring

The IoT plays a key role in enabling real-time monitoring of machines and equipment used in various industries. By embedding smart sensors into physical assets—such as motors, turbines, or pumps—important operating parameters like temperature, vibration, current, and pressure can be continuously tracked. These sensors collect data at regular intervals and transmit it to a central system, allowing maintenance teams to observe machine conditions without the need for manual inspection.

With IoT, this monitoring becomes proactive. If a machine shows early signs of abnormal behavior, such as unusual vibrations or overheating, the system can issue warnings before a failure occurs. This helps in taking timely maintenance actions, reducing unplanned downtime and minimizing repair costs. IoT devices are essential for collecting large amounts of real-world data, which are later used to train AI models for predicting equipment health.

3.7.2 Edge computing for local AI processing

Edge computing involves processing data near the source where it is generated—typically on or near the machines themselves—instead of sending all data to a remote cloud server. This approach is especially useful in time-sensitive applications like PdM, where quick decision-making is critical.

By using edge devices with built-in AI capabilities, some analysis can be done locally. For example, a smart sensor attached to a motor can run a lightweight ML model to detect abnormal patterns and trigger alerts instantly. This reduces the dependency on internet connectivity and cuts down the time taken to respond to potential issues.

Edge computing also minimizes the amount of data that needs to be sent over networks, which is helpful in environments with limited bandwidth or high data generation rates. It adds a layer of speed and resilience to the PdM system.

3.7.3 Cloud platforms for data storage and analytics

Cloud computing provides a scalable and flexible platform for handling large volumes of data collected from IoT devices. Cloud platforms can store data from hundreds or even thousands of sensors across multiple locations, allowing centralized access for analysis and reporting.

In PdM systems, the cloud is often used to host advanced AI models, historical data records, and visualization dashboards. These resources help engineers analyze long-term trends, train more accurate prediction models, and coordinate maintenance activities across different sites.

Moreover, cloud platforms support remote access, which is valuable for teams that need to monitor equipment from different geographical locations. They also offer integration with various data analysis and ML tools, making it easier to update models and scale the system as more devices are added.

3.7.4 Cybersecurity and data privacy concerns

As PdM systems rely heavily on interconnected devices and online platforms, they become vulnerable to cyberthreats. Unauthorized access, data tampering, or malware attacks can compromise the integrity of the data and the reliability of the maintenance system.

To address these concerns, strong cybersecurity measures are essential. This includes encrypting data during transmission and storage, using secure authentication methods, and regularly updating software to patch vulnerabilities. Devices must also be protected against physical tampering.

Data privacy is another important issue, especially when sensitive operational data is shared over networks or stored in third-party cloud systems. Organizations must ensure compliance with data protection regulations and restrict access to only authorized personnel.

Maintaining trust in the system requires a balance between connectivity and security, making cybersecurity a core component of any IoT–AI-integrated maintenance framework.

3.8 CASE STUDIES AND APPLICATIONS

3.8.1 Predictive maintenance in wind turbines

Wind turbines operate under varying environmental conditions and are often installed in remote locations, making regular physical inspections difficult and costly. To address this, PdM using AI has been introduced to monitor the health of key components such as gearboxes, blades, and generators. Sensors embedded in the turbine continuously gather data such as vibration levels, temperature, rotational speed, and electrical output. AI algorithms analyze this data to detect early signs of wear or imbalance.

For instance, vibration data can reveal changes in the gear meshing frequency, which may indicate bearing wear or gear tooth damage. ML models trained on historical fault data can predict the likelihood of component failure, enabling timely interventions. This approach reduces unscheduled downtime and extends the operational life of turbines while lowering maintenance costs.

3.8.2 Fault prediction in power transformers

Power transformers are vital to electricity transmission and distribution, and their failure can cause major blackouts and financial losses. Traditional maintenance often misses early-stage faults. PdM supported by AI addresses this issue by analyzing data such as oil temperature, dissolved gas concentrations, load cycles, and partial discharge signals.

Using pattern recognition techniques, AI can identify abnormal trends in the transformer's behavior. For example, an increase in specific gases like acetylene or ethylene in transformer oil may indicate internal arcing or overheating. Deep learning models like LSTM networks are particularly useful for time-series analysis of such sensor data, allowing the system to learn patterns over time and forecast failures well in advance.

3.8.3 Maintenance of industrial motors and pumps

Motors and pumps are core components in manufacturing plants, water treatment facilities, and power stations. PdM for these machines focuses on identifying early signs of mechanical or electrical issues such as misalignment, unbalance, rotor faults, and cavitation.

Condition monitoring devices collect data including current consumption, vibration, acoustic signals, and temperature. AI systems process this information to detect deviations from normal operation. For example, an increase in current may indicate increased mechanical load due to worn bearings. Classification algorithms like SVMs or neural networks can label these patterns as potential faults, prompting maintenance teams to take preemptive action.

This not only prevents unexpected breakdowns but also improves energy efficiency by ensuring that machines operate under optimal conditions.

3.8.4 Real-world implementation challenges and outcomes

While the benefits of AI-based PdM are significant, real-world implementation is not without challenges. One major issue is **data quality**—incomplete, noisy, or inconsistent data can lead to inaccurate predictions. Ensuring reliable sensor networks and data collection protocols is critical.

Another challenge is the **integration of AI systems** with existing industrial setups. Many facilities operate with legacy equipment that lacks modern communication interfaces, making it difficult to deploy AI models seamlessly. Moreover, the **interpretability** of AI models is often a concern; maintenance personnel may hesitate to trust black-box predictions without clear explanations.

Despite these hurdles, companies that have successfully adopted PdM report clear benefits: fewer equipment failures, reduced maintenance costs, and improved operational efficiency. In the wind energy sector, for example, PdM has led to increased availability and performance of turbines. Similarly, power utilities have reduced transformer outages and extended their service life.

3.9 CHALLENGES AND LIMITATIONS

3.9.1 Data availability and quality issues

AI models depend heavily on large volumes of high-quality data to perform effectively. In many industries, such data may not be readily available. Older equipment may lack sensors or digital monitoring systems, making data collection difficult. Even when data is available, it may be incomplete, noisy, or inconsistent due to sensor errors, manual logging, or environmental conditions. Low-quality data can mislead the AI model, resulting in inaccurate predictions and false alarms, which reduces trust in the system. Proper data preprocessing, filtering, and validation are essential but often require additional resources and expertise.

3.9.2 High initial cost and infrastructure requirements

Implementing AI-based PdM involves significant upfront investment. This includes the cost of installing sensors and monitoring equipment, setting up data storage systems, and investing in computing infrastructure for training and running AI models. In some cases, companies also need to hire or train skilled personnel to manage and interpret the results. These requirements can be a barrier for small and medium-sized enterprises, which may not have the capital or resources to adopt such technology despite the long-term savings it promises.

3.9.3 Interpretability of AI models

Many AI models, especially those based on deep learning, are often viewed as "black boxes" because their decision-making processes are not easily understood. This lack of transparency can be problematic in maintenance environments where engineers and technicians need to understand the reasons behind a fault prediction. Without clear explanations, it becomes difficult to trust the model's output or take appropriate action. This challenge has led to increased interest in developing interpretable or explainable AI models, but such solutions are still under active research and development.

3.9.4 Resistance to technology adoption in traditional industries

In industries that have relied on conventional maintenance methods for decades, there can be hesitation or resistance to adopting AI-based systems. This reluctance may stem from concerns about job displacement, uncertainty about the effectiveness of new technology, or simply a lack of familiarity. Changing established workflows and retraining personnel require time, commitment, and often a cultural shift within the organization. Overcoming this resistance involves clear communication of benefits, hands-on training, and gradual integration of AI tools into existing processes.

3.10 FUTURE TRENDS AND RESEARCH DIRECTIONS

3.10.1 Self-learning maintenance systems

The future of PdM lies in systems that can learn and adapt without constant human intervention. Self-learning maintenance systems utilize feedback from real-world operations to improve their predictive accuracy over time. Unlike static models that require manual retraining, these systems update themselves as new data is collected. They can detect changes in equipment behavior due to aging, wear, or changing environmental conditions and adjust their predictions accordingly. This leads to more precise forecasting and reduced false alarms, helping industries to achieve a higher level of automation and reliability.

3.10.2 Explainable AI (XAI) for maintenance decision-making

While traditional AI models can offer high prediction accuracy, many function as "black boxes," making it difficult for engineers and technicians to understand how decisions are made. XAI addresses this issue by providing insights into the reasoning behind model predictions. In the context of PdM, XAI helps maintenance teams understand why a certain failure is expected, which data points contributed most to the prediction, and how to prioritize corrective actions. This transparency builds trust in AI systems and supports informed decision-making, especially in safety-critical industries like energy, aviation, and manufacturing.

3.10.3 Integration with Industry 4.0 and smart factories

PdM is a core element of Industry 4.0, which promotes the use of digital technologies to create intelligent, interconnected, and flexible production environments. In smart factories, machines, sensors, and AI systems work together to monitor equipment health in real time. The integration of AI with industrial communication networks allows PdM systems to coordinate with other systems, such as inventory management or scheduling software. This leads to more efficient use of resources, reduced production delays, and quicker response to potential failures. As industries embrace full digital transformation, PdM will become a standard part of the manufacturing ecosystem.

3.10.4 Potential of AI in sustainable and renewable energy maintenance

AI has a significant role to play in supporting the transition to renewable energy sources. Wind turbines, solar panels, and other clean energy systems often operate in remote or harsh environments, making manual inspection difficult and costly. AI-powered PdM can monitor these systems continuously using data from sensors and drones, predicting faults before they cause energy losses. This not only increases the reliability of energy production but also extends the lifespan of renewable energy equipment. Furthermore, AI can help optimize maintenance schedules to reduce unnecessary site visits, minimizing carbon footprints and supporting the sustainability goals of energy providers.

3.11 CONCLUSION

3.11.1 Summary of key AI techniques in predictive maintenance

In this chapter, we have explored how various AI techniques are being applied to improve PdM strategies across industries. Techniques such as **supervised learning** (like decision trees and SVMs), **unsupervised learning** (such as clustering and dimensionality reduction), and **deep learning methods** (including CNNs, RNNs, and LSTMs) play a crucial role in analyzing data collected from machines and equipment. These techniques help detect early signs of wear, identify fault patterns, and forecast equipment failures before they occur.

Additionally, **hybrid approaches** that combine different AI models have shown improved accuracy and adaptability. AI systems enhanced by **fuzzy logic, expert systems,** or **ensemble models** offer better flexibility and decision-making under uncertain or incomplete conditions. With the aid of **IoT devices**, data can be continuously gathered in real time, enabling AI models to perform live monitoring and dynamic analysis of equipment health.

3.11.2 Overall impact on reliability and efficiency

By using AI in PdM, industries benefit from **increased system reliability**, as potential failures are detected early and addressed before they cause breakdowns. This leads to **reduced unplanned downtime**, which saves both time and money. Maintenance can be planned more effectively, resources can be better allocated, and unnecessary part replacements can be avoided.

Furthermore, the ability of AI to process large volumes of data with high speed and precision improves decision-making and helps operators focus on the most critical tasks. Over time, this leads to **longer equipment life**, **lower maintenance costs**, and **safer working conditions**, especially in critical sectors like energy, manufacturing, and transportation.

3.11.3 The way forward for research and industrial practice

Looking ahead, the integration of AI in PdM is expected to become even more intelligent and autonomous. **Future research** should focus on improving model accuracy with less training data, enhancing model interpretability, and developing **explainable AI** tools to build trust among engineers and decision-makers.

There is also a need to address challenges such as **data security, privacy**, and the **integration of legacy systems** into modern AI-based platforms. Developing lightweight AI models suitable for **edge devices** and creating **standardized frameworks** for deploying AI in different industrial environments will be crucial steps.

In industrial practice, companies should aim to build **data-driven maintenance cultures**, invest in **training personnel**, and gradually implement AI systems in a scalable and sustainable manner. As technology advances, PdM powered by AI will not just be a competitive advantage but a core requirement for efficient and future-ready operations.

REFERENCES

1. Jardine, A. K. S., Lin, D., & Banjevic, D. (2006). A review on machinery diagnostics and prognostics implementing condition-based maintenance. *Mechanical Systems and Signal Processing*, 20(7), 1483–1510.
2. Lei, Y., Li, N., Guo, L., Li, N., Yan, T., & Lin, J. (2020). Machinery health prognostics: A systematic review from data acquisition to RUL prediction. *Mechanical Systems and Signal Processing*, 104761.

3. Widodo, A., & Yang, B. S. (2007). Support vector machine in machine condition monitoring and fault diagnosis. *Mechanical Systems and Signal Processing*, 21(6), 2560–2574.
4. Zhang, W., Yang, D., & Wang, H. (2019). Data-driven methods for predictive maintenance of industrial equipment: A survey. *IEEE Systems Journal*, 13(3), 2213–2227.
5. Si, X. S., Wang, W., Hu, C. H., & Zhou, D. H. (2011). Remaining useful life estimation–A review on the statistical data driven approaches. *European Journal of Operational Research*, 213(1), 1–14.
6. Hodge, V. J., O'Keefe, S., Austin, J., & Polovina, S. (2016). A survey of outlier detection methodologies. *Artificial Intelligence Review*, 22(2), 85–126.
7. Wang, T., Yu, J., & Siegel, D. (2008). A similarity-based prognostics approach for remaining useful life estimation of engineered systems. In: *IEEE International Conference on Prognostics and Health Management*, Denver, CO.
8. Ghosh, S., Kumar, S., & Verma, M. (2020). Predictive maintenance using machine learning: A survey. *International Journal of Computer Applications*, 59, 559–573.
9. Jayasinghe, L., & Amarasinghe, M. (2020). AI-powered predictive maintenance for industrial IoT: A review. *Journal of Industrial Information Integration*, 17, 100123.
10. Wu, B., & Luo, Y. (2021). Condition monitoring and fault diagnosis using deep learning. *Applied Sciences*, 11(7), 3056.
11. Goodfellow, I., Bengio, Y., & Courville, A. (2016). *Deep Learning*. MIT Press.
12. Bishop, C. M. (2006). *Pattern Recognition and Machine Learning*. Springer.
13. Liu, R., Yang, B., Zio, E., & Chen, X. (2018). Artificial intelligence for fault diagnosis of rotating machinery: A review. *Mechanical Systems and Signal Processing*, 108, 33–47.
14. Qin, Y., & Li, J. (2020). Industrial predictive maintenance with deep learning: A review. *Journal of Intelligent Manufacturing*, 31, 433–452.
15. Zhang, R., & Wang, L. (2018). Fault diagnosis of rotating machinery using a deep convolutional neural network. *Computers in Industry*, 107, 29–40.
16. Yan, R., Gao, R. X., & Chen, X. (2014). Wavelets for fault diagnosis of rotary machines: A review with applications. *Signal Processing*, 96, 1–15.
17. Duan, Y., Street, W. N., & Xu, Y. (2013). A survey of hybrid recommendation techniques. *ACM Computing Surveys (CSUR)*, 45(4), 1–33.
18. Qin, C., & Pan, E. (2019). Machine learning for fault diagnosis of rotating machinery: An overview. *Advances in Mechanical Engineering*, 11(5), 1–12.
19. Mobley, R. K. (2002). *An Introduction to Predictive Maintenance*. Butterworth-Heinemann.
20. Längkvist, M., Karlsson, L., & Loutfi, A. (2014). A review of unsupervised feature learning and deep learning for time-series modeling. *Pattern Recognition Letters*, 42, 11–24.
21. Singh, P., Arora, K., Rathore, U. C., Joshi, G. P., & Cho, W. (2024). Comparative study of controllers in battery energy storage system integrated with doubly fed induction generator-based wind energy conversion system for power quality improvement. *Energy Reports*, 11(9), 4587–4600. https://doi.org/10.1016/j.egyr.2024.04.020.

22. Vapnik, V. N. (1995). *The Nature of Statistical Learning Theory*. Springer.
23. Zhang, C., Li, Y., & Zhou, D. (2019). IoT and AI in predictive maintenance for smart manufacturing. *Journal of Manufacturing Systems*, 51, 22–33.
24. Mhamdi, M., & Mahjoub, M. A. (2021). A survey on AI-based maintenance strategies in the industry. *Procedia Computer Science*, 194, 144–153.
25. Manickam, P., Girija, M., Dutta, A. K., Ramesh Babu, P., Arora, K., Jeong, M. K., & Acharya, S. (2024). Empowering cybersecurity using enhanced rat swarm optimization with deep stack-based ensemble learning approach. *IEEE Access*. https://doi.org/10.1109/ACCESS.2024.3395328.
26. Lee, J., Bagheri, B., & Kao, H. A. (2015). A cyber-physical systems architecture for industry 4.0-based manufacturing systems. *Manufacturing Letters*, 3, 18–23.
27. Nair, S., & Narayan, A. (2020). Predictive maintenance using deep learning: A case study. *International Journal of Prognostics and Health Management*.
28. Zhao, R., Yan, R., Chen, Z., Mao, K., Wang, P., & Gao, R. X. (2019). Deep learning and its applications to machine health monitoring. *Mechanical Systems and Signal Processing*, 115, 213–237.
29. Qiu, H., Lee, J., Lin, J., & Yu, G. (2006). Wavelet filter-based weak signature detection method and its application on rolling element bearing prognostics. *Journal of Sound and Vibration*, 289(4–5), 1066–1090.
30. Wang, L. (2020). Intelligent predictive maintenance for industry 4.0 equipment using machine learning. *Procedia Manufacturing*, 42, 478–485.
31. Cheng, S., Azarian, M. H., & Pecht, M. (2010). Sensor systems for prognostics and health management. *Sensors*, 10(6), 5774–5797.
32. Bhatt, B., Sharma, H., Arora, K., Joshi, G. P., & Shrestha, B. (2023). Levy flight-based improved grey wolf optimization: A solution for various engineering problems. *Mathematics*, *11*(7), 1745. https://doi.org/10.3390/math11071745.
33. O'Donovan, P., Leahy, K., Bruton, K., & O'Sullivan, D. T. (2015). An industrial big data pipeline for data-driven analytics maintenance applications in large-scale smart manufacturing facilities. *Journal of Big Data*, 2(1), 1–26.
34. Zhao, M., Kang, M., & Li, Y. (2017). Deep residual networks with dynamically weighted wavelet coefficients for fault diagnosis of rotating machinery. *IEEE Transactions on Industrial Electronics*, 65(2), 1503–1512.
35. Khan, S., Yairi, T., & Seung, H. (2021). Machine learning for fault detection and predictive maintenance. *Computers & Industrial Engineering*, 149, 106889.
36. Schmidt, D. C., & Austin, J. (2016). Deep learning for industrial AI: A case study. *Industrial AI Journal*, 1(1), 23–34.
37. Wang, P., & Shao, H. (2019). Intelligent fault diagnosis using deep transfer learning. *Reliability Engineering & System Safety*, 184, 55–66.
38. Patel, H., & Shah, D. (2020). Predictive maintenance using IoT and machine learning. *International Journal of Engineering Research and Technology*, 9(5), 654–659.
39. Das, R., & Cheng, Q. (2019). Data preprocessing for machine learning in predictive maintenance. *Data Science Journal*, 18(1), 1–15.
40. Babu, G. S., Zhao, P., & Li, X. L. (2016). Deep convolutional neural network-based regression approach for estimation of remaining useful life. *International Conference on Database Systems for Advanced Applications*.

41. Heimes, F. O. (2008). Recurrent neural networks for remaining useful life estimation. *IEEE International Conference on Prognostics and Health Management.*
42. Li, X., Ding, Q., & Sun, J. Q. (2018). Remaining useful life estimation in prognostics using deep convolution neural networks. *Reliability Engineering & System Safety*, 172, 1–11.
43. Bousdekis, A., Mentzas, G., & Apostolou, D. (2015). Predictive maintenance using machine learning: A literature review. *International Journal of Computer Applications*, 132(12), 1–6.
44. Zhang, Y., Qiu, T., & Wu, D. (2019). A survey of machine learning for big data processing. *EURASIP Journal on Advances in Signal Processing*, 2019(1), 1–16.
45. Tsui, K. L., Chen, N., Zhou, Q., Hai, Y., & Wang, W. (2015). Prognostics and health management: A review on data-driven approaches. *Mathematical Problems in Engineering*, 2015.
46. Arora, K., Joshi, G. P., Ragab, M., Rawa, M., Milyani, A. H., Mansour, R. F., & Yang, E. (2022). Bilateral contract for load frequency and renewable energy sources using advanced controller. *Computers, Materials & Continua*, 73(2), 3165–3180. https://doi.org/10.32604/cmc.2022.026966.
47. Wang, Y., & Yu, L. (2021). Real-time predictive maintenance using neural networks. *IEEE Access*, 9, 98765–98775.
48. Arora, K., Kumar, A., Kamboj, V. K., Prashar, D., Jha, S., Shrestha, B., & Joshi, G. P. (2020). Optimization methodologies and testing on standard benchmark functions of load frequency control for interconnected multi-area power system in smart grids. *Mathematics*, 8(6), 980. https://doi.org/10.3390/math8060980.
49. Wang, S., Wang, X., & Wu, Y. (2020). Predictive maintenance model using CNN and LSTM for industrial equipment. *Applied Sciences*, 10(4), 1453.
50. ISO 17359:2018. (2018). *Condition Monitoring and Diagnostics of Machines: General Guidelines*. International Organization for Standardization.

Chapter 4

Artificial intelligence in additive manufacturing

A comprehensive review of applications, challenges, and future directions

Ashish Khaira and Neeraj Chandnani

4.1 INTRODUCTION

Additive manufacturing (AM), commonly known as 3D printing, is an innovative technology in modern manufacturing, building objects layer by layer from ordinary models. It has many more benefits than traditional manufacturing techniques, as well as the ability to create complex geometries, reducing waste, and the ability to swiftly sample and manufacture parts. Due to its variability and efficiency, AM is taking center lane in an array of areas, such as aerospace, medicine, and automotive.

AM is revolutionized by artificial intelligence (AI) because AI can handle the complexity of the AM process and enable real-time decision-making. ML and DL—two popular approaches of AI—have the ability to maximize process parameters, predict and monitor part quality, and detect defects before the production of the part is completed. AI incorporation makes AM processes more effective, consistent, and with the ability to produce high-quality parts with minimum human involvement.

4.1.1 Scope

Although AM techniques and AI methods can be synergizing toward complementary solutions, AI methods in AM are covered in the paper via different case studies, challenges in the integration of both technologies, and opportunities. It covers several AI techniques, including ML and DL, and explains their applications in process optimization, design innovation, and quality assurance. The review also covers technical and operational aspects of AI implementation in AM, together with trends and research directions for the future.

4.1.2 Objective

The purpose of this research is to integrate prevailing research directions in the confluence of AI and AM, determine areas of gap in the prevailing literature, and suggest future research directions on how to apply AI in pushing the future of AM. Through presenting a thorough review, this report seeks

 DOI: 10.1201/9781003641582-4

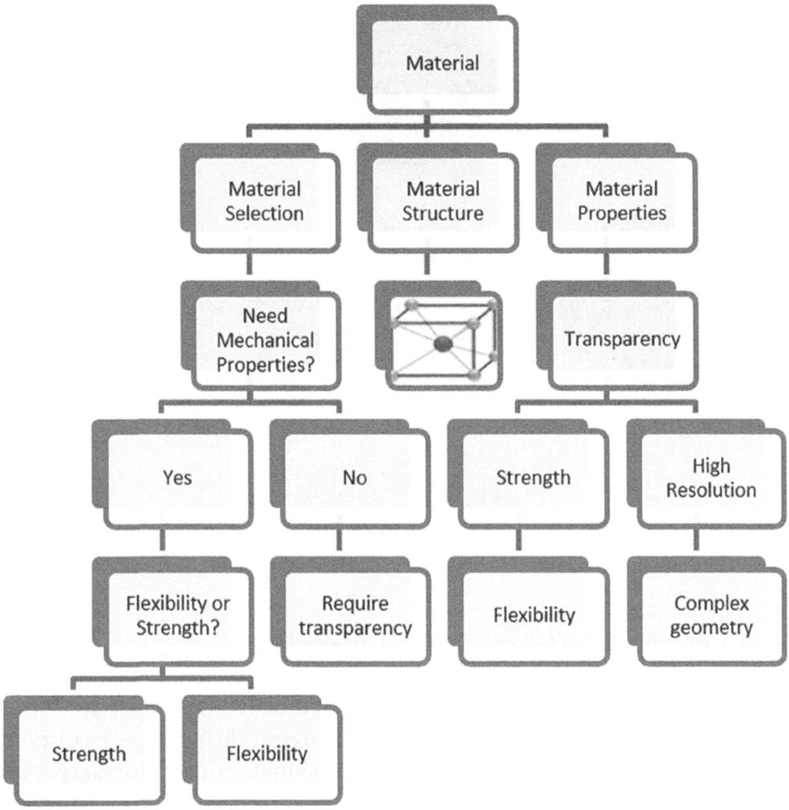

Figure 4.1 AI-based material creation for the AM process.

to underscore the potential for AI in transforming AM and provide insight into how these technologies can be harnessed and integrated to augment manufacturing processes [1–4].

Figure 4.1 demonstrates the synergy between AM and machine learning (ML) for material development, targeting three main aspects: material selection, material structure, and material properties. Figure 4.1 depicts a systematic and intelligent process of material design.

4.2 AI METHODOLOGIES IN AM

4.2.1 Machine learning (ML)

4.2.1.1 Techniques

Supervised learning: Methods like artificial neural networks (ANNs) and support vector machines (SVMs) are applied for parameter

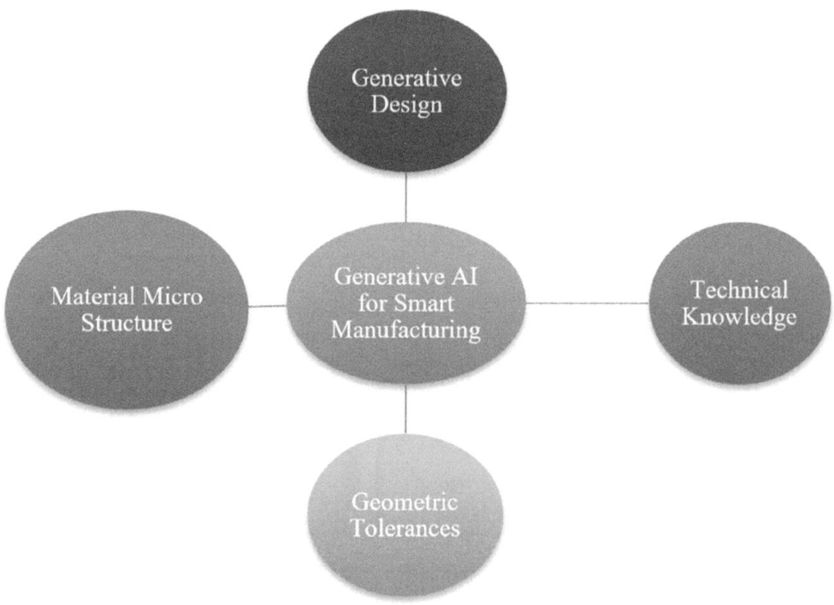

Figure 4.2 AI-aided design for additive manufacturing (AM).

optimization in AM. These approaches are utilized for predicting and regulating the quality of the parts by optimizing the process parameters.

Unsupervised learning: Clustering algorithms are used for defect identification in AM processes. The algorithms are able to detect patterns and anomalies in the data, which are a sign of defects, without labeled training data.

Figure 4.2 demonstrates the idea of AI-aided design for AM. The theme is a ML model (represented with a diagram of a neural network) that uses different data inputs to improve design processes in AM. The picture demonstrates how AI combines data-driven design, expert knowledge, structural optimization, and correction for real-world deviations to improve AM. It's a closed-loop system where every element provides input to the AI model to generate wiser, more accurate designs.

Case study 4.1: Bayesian ML for cross-printer knowledge transfer

One of the most interesting claims of ML in AM is the design of a framework for effectively transferring metal AM knowledge between various 3D printers. This method is an efficient solution for the large amount of time and money required to optimize AM processes for new printers. The authors suggest the application of data mining methods to gather and analyze available AM data on Ti-6Al-4V and then apply Bayesian ML models to forecast optimal process parameters for new printers. The approach dramatically outperforms other models such as SVMs and logistic regression. The framework is verified by experiments on three scenarios of real-world adoption challenges with different levels of technological similarity between printers. The end vision is to drive the transition to new metals' AM technologies through existing knowledge and reducing the amount of trial-and-error experimentation that must be performed [5].

4.2.2 Deep learning (DL)

4.2.2.1 *Techniques*

- **Convolutional Neural Networks (CNNs)**: They are utilized for *in situ* monitoring in AM, for instance, the identification of porosity in real time during printing. CNNs may be employed to analyze images taken during manufacturing for detecting defects and guaranteeing quality in printed parts.
- **Generative Adversarial Networks (GANs)**: GANs are used in generative design, such as topology optimization. They are capable of creating innovative design solutions with optimized material utilization and structural efficiency, beyond what conventional design processes can achieve.

Case study 4.2: modified DenseNet (NDenseNet) for microstructure classification

A new DL approach has been created to assess material performance in laser AM. The authors utilize an adapted DenseNet CNN, named NDenseNet, to categorize microstructure images of Ti6-Al-4V alloy with 90.4% accuracy. Major enhancements to the NDenseNet model are a new "mix-pooling" method and group convolutions, leading to lower computational expense and higher accuracy than other models such as ResNet and DenseNet. In addition, transmission learning from a breast histology image database improves the presentation of the model, showing an industrial application of DL to speed up material classification [6].

4.2.3 Hybrid AI approaches

4.2.3.1 *Genetic algorithms (GAs) for multiobjective optimization*

The use of GAs for multiobjective optimization in AM GAs resembles natural selection in order to find optimal solutions to the problems with multiple objectives. As an example in two-photon lithography (TPL), GAs can be applied for the optimization of various parameters throughout the various process parameters, such as light dosage, resin type, and structure design to achieve the required quality and performance of printed components. This approach enables the design of ML and computer vision algorithms for the real-time monitoring and optimization of the process in AM [1].

4.2.3.2 Physics-informed ML integrating thermal/mechanical laws

Physics-informed ML combines thermal and mechanical concepts to improve the accuracy and reliability of predictions in the AM process. By incorporating underlying physics principles, models can better capture and predict material behavior for various manufacturing conditions [7]. Physics-informed ML overcomes issues related to anisotropic material behavior and intricate interactions between process parameters and product quality. It also speeds up material discovery and design, enabling more intelligent and efficient AM processing. Data-driven integrated frameworks play a key role in making the most out of these advances and pushing AM [4].

4.3 AI APPLICATIONS IN AM

4.3.1 Process optimization

4.3.1.1 Parameter tuning

The parameter optimization is a vital aspect of process optimization in AM. Process tuning like laser power, scanning speed, and other process parameters can have a material impact on the functionality and quality of the parts being produced. For example, in TPL, photo-curable resins coupled with modifications in light dosages allow the recording of the 3D printing process, thus providing a complete sweep of parameters. The data allows the development of ML and computer vision algorithms for real-time process monitoring and optimization in AM [1]. Additionally, an ML system has been implemented to successfully transfer knowledge about metal AM across different 3D printers. Using Bayesian ML models, the system estimates suitable process parameters for new printers and minimizes the time and cost incurred in process optimization by a significant extent [5]. A systematic review of the use of AI in AM reports the widespread use of AI methodologies for process optimization, notably in powder bed fusion (PBF) and fused deposition modeling (FDM), using different AI methodologies like CNNs and ANNs [2].

4.3.1.2 Real-time closed-loop control

Real-time closed-loop control is a fundamental aspect of process optimization in AM. This technique involves ongoing monitoring of the manufacturing process and instant application of corrections in order to maintain the conditions optimized. For example, in melt electrowriting, the real-time closed-loop regulator can be used to facilitate precise control of the deposition process, thereby guaranteeing the creation of first-class printed parts [8]. A digital and scalable AM platform exploits the convergence of cyber-physical systems and digital twins to connect data frameworks and manufacturing equipment. The Augmented Intelligence Relay (AIR) applies a sequence of AI algorithms integrated with human knowledge to predict component performance and reduce reliance on large-scale physical testing, thus enhancing and combining the AM process through instance-specific, data-driven part qualification [9].

4.3.2 Design and material innovation

4.3.2.1 Generative design for lightweight structures

Generative design is among the most significant functions in AM for lightweight structure design [10]. A good example is the generation and production of the steering upright of the Formula SAE race car with minimal weight. The approach used is topology optimization, a computational algorithm for weight reduction with strict structural constraints defined using telemetry data and finite element analysis (FEA) of a baseline component [11]. The optimized model generated from the process has features of design for additive manufacturing (DfAM) and Design for Finishing (DfF) and resulted in a 9% weight loss from the baseline component. The manufactured part was validated by FEA and successfully produced by laser PBF, demonstrating a complete methodology combining structural analysis, manufacturing factors, and iterative design optimization to create a manufacturable and competitive component [12].

4.3.2.2 ML-driven material discovery

ML is important in material discovery for AM. A new DL technique, for example, uses a variant of DenseNet CNN, referred to as NDenseNet, to recognize microstructure images of Ti6-Al-4V alloy with 90.4% accuracy. Most important enhancements in the NDenseNet model are a new "mix-pooling" method and application of group convolutions, which lead to a lower computational cost and higher accuracy than other models such as ResNet and DenseNet [13]. Transfer learning from the breast histology image dataset still boosts the performance of the model, presenting an actual application of DL for speeding up material characterization in industry [6].

Furthermore, an ML framework has been implemented to effectively transfer knowledge regarding metal AM from one 3D printer to another. Utilizing Bayesian ML models, this framework estimates optimal process parameters for new printers and saves considerable time and cost of process optimization. This method exploits available AM data on Ti-6Al-4V and reduces the amount of extensive trial-and-error experiments, facilitating rapid adoption of new metals' AM technologies [5].

4.3.3 Quality assurance

4.3.3.1 Defect detection via in situ imaging (CNNs)

Defect inspection is a quality control feature of AM. CNNs are used mainly for in situ imaging for real-time defect detection during printing. For instance, in TPL, videos of varied dosages of light and photo-curable resins capture the 3D printing process, showing a big parameter sweep. Unprocessed videos and cropped frames of the videos, where points of interest of the region of interest around the printed object have been labeled, are included in the dataset. The resource is intended to be used with ML and computer vision algorithm development, specifically object segmentation, localization, and inspection for part quality in TPL. The extensive collection of TPL processes in various polymerization states (uncured, cured, and damaged) is ideal to train and validate algorithms to be used in monitoring and optimizing processes in real time in AM [1].

4.3.3.2 Fatigue life prediction (RF/ANN models for 316L steel)

Fatigue life prediction of components produced by AM is crucial for their long-term reliability and service life performance. ML models like Random Forests (RFs) and ANNs are employed to estimate the fatigue life of 316L stainless steel components. Scientists have developed a platform wherein continuum damage mechanics (CDM)-based fatigue life predictions are integrated with three ML algorithms: ANNs, RFs, and SVMs. Equations developed from the CDM account for several AM process parameters, such as hatch space, laser power, scan speed, and powder layer thickness, and fatigue loading conditions [14]. Experimental data is compared with predictions of the ML models in the present work, with RF being the best choice. Parametric studies are performed to optimize the performance of the ML models and study the effect of varying parameters on fatigue life [15].

4.3.4 Sustainability

4.3.4.1 AI for circular economy (e.g., repair/restoration frameworks)

AI has an important influence on optimizing component design for AM repair and renovation in a circular economy context. This method differentiates traditional remanufacturing processes from the automated processes of AM, accentuating the merits and demerits of each. The focus is primarily on using AI, namely GAs, to optimize component designs to ensure efficient and sustainable AM-based repair. This identifies gaps in existing research and suggests an alternative framework for upcoming studies. Combining design aspects with AM processes enhances efficiency, quality, and environmental benefits in remanufacturing [16].

4.3.4.2 Energy-efficient AM workflows enabled by AI

Generative artificial intelligence (generative AI) holds great promise to drive the sustainability agenda of Industry 5.0 manufacturing. Through a mixed-methods research design involving case studies and interviews, researchers have created a strategic roadmap for ten critical generative AI functions. These range from sophisticated customer support to operational resilience and, synergistically, add up to more than the sum of their individual contributions toward sustainability [17]. The research questions make use of Interpretive Structural Modeling (ISM) and MICMAC analysis to depict the interdependence and hierarchical structure of these functions, serving as a pragmatic guideline for manufacturers to consider generative AI integration strategically. The research endeavors to link the gap among Industry 5.0 theoretical ideals and their real-world application through generative AI, generating theoretical and practical implications for sustainable manufacturing [18].

4.4 INDUSTRY-SPECIFIC IMPLEMENTATIONS

4.4.1 Aerospace

- **AI-Optimized Topology Designs (e.g., Turbine Blades):** This review article in 2015 analyzes the status quo and future scope of AM in metal components within the aerospace sector. Electron beam melting (EBM), selective laser melting (SLM), and other metal deposition techniques are considered by the authors as the most viable AM methodologies for aerospace as they point to the need for precise process modeling so as to be able to predict material properties as well as microstructure. A key challenge addressed is the need for rigorous certification and quality control to meet aerospace safety regulations, emphasizing the need for further standardization and model development to link the gap among AM's potential and its real-world application in the aerospace sector. The article also reviews various alloys used in AM, particularly titanium and nickel-based alloys, and analyzes the relationship between processing parameters and resulting material properties like fatigue behavior and surface finish.
- **Certification Challenges and Digital Twin Solutions:** This research paper details a scalable digital platform for AM, addressing the

current limitations of inconsistent part properties and high variability in AM processes. The platform leverages digital twins, constantly updated computer models of physical parts, enabled by a cyberphysical infrastructure connecting manufacturing equipment and data systems. A key innovation is the AIR, which uses a series of AI algorithms and human expertise to predict component performance with less reliance on extensive physical testing [19]. The ultimate goal is to create a more efficient and reliable AM process through data-driven, instance-specific part qualification (Paragraph 16).

4.4.2 Healthcare

- **4D Printing for Bioengineered Implants**: This article explores the synergistic potential of combining AM, specifically 4D printing, and AI to design advanced intelligent systems. The authors highlight how 4D printing, which incorporates a time-dependent element allowing for shape and property changes, expands design possibilities beyond traditional manufacturing limitations. AI's role is crucial in overcoming the computational challenges associated with complex designs and material properties, optimizing designs, and predicting material behavior. The integration of AM and AI is presented as a pathway to "5D printing," where AI provides real-time control and optimization during the manufacturing process, leading to the creation of multifunctional and self-adapting systems with applications spanning robotics and bioengineering (Paragraph 15).
- **ML in Biomanufacturing (e.g., Scaffold Design)**: This research paper from the *International Journal of Chemistry, Mathematics, and Physics* explores the transformative impact of AI on biotechnology manufacturing. The authors investigate how AI enhances various aspects of biomanufacturing, including process optimization, predictive maintenance, quality control, and supply chain management. The paper details current manufacturing technologies in biotechnology, focusing on the shift toward continuous processing and automation, and then examines specific AI applications, highlighting both their benefits and the associated challenges like data privacy and workforce retraining needs. Ultimately, the study emphasizes AI's potential to revolutionize biotechnology manufacturing, leading to increased efficiency, reduced costs, higher product quality, and accelerated innovation (Paragraph 4).

4.4.3 Automotive

- **Generative AI for Part Consolidation (Weight Reduction)**: This research article details the design and AM of a lighter Formula SAE racing car steering upright. The process utilizes topology optimization, a numerical method, to minimize weight while meeting strict structural constraints determined from telemetry data and FEA of a reference component. The optimized design, incorporating principles of DfAM and DfF, resulting in a 9% weight reduction compared to the original, and the final product underwent FEA validation and successful manufacturing via laser PBF. The study emphasizes a holistic approach, integrating structural analysis, manufacturing considerations, and iterative design refinement to achieve a competitive and manufacturable part (Paragraph 2).

4.5 CHALLENGES

4.5.1 Technical

- **Data Scarcity and the Need for Synthetic Datasets (GANs)**: This data article describes a dataset of videos showcasing TPL, an AM technique. The videos, taken under varying light dosages, photo-curable resins, and structure types, capture the 3D printing process, offering a comprehensive parameter sweep. The dataset includes both raw videos and cropped video frames focused on regions of interest around printed parts, along with a Python script for reproducing the organized directory of cropped frames. This resource aims to facilitate ML and computer vision algorithm development, particularly for applications in object segmentation, localization, and part quality assessment in TPL. The data's value lies in providing a rare and extensive collection of TPL processes exhibiting different polymerization states (uncured, cured, and damaged) suitable for training and validating algorithms for real-time process monitoring and optimization in AM (Paragraph 1).
- **Model Interpretability (XAI Frameworks)**: This academic review article explores the application of data-driven approaches, particularly ML, to revolutionize AM. It examines how these methods can overcome AM's challenges, such as anisotropic material properties and the complex relationship between process parameters and product quality, by accelerating material discovery and design. The authors detail the use of data-driven techniques across various AM stages, from material selection (homogeneous and composite materials) and structural design (topology optimization, generative design) to tool path optimization and process parameter tuning. Ultimately, the paper advocates for the development of integrated data-driven frameworks to enable a "smarter" and more efficient AM process, emphasizing the need for comprehensive databases to facilitate this progress (Paragraph 17).

4.5.2 Operational

- **Integration with Legacy Systems**: This research paper details a scalable digital platform for AM, addressing the current limitations of inconsistent part properties and high variability in AM processes. The platform leverages digital twins, constantly updated computer models of physical parts, enabled by a cyberphysical infrastructure connecting manufacturing equipment and data systems. A key innovation is the AIR, which uses a series of AI algorithms and human expertise to predict component performance with less reliance on extensive physical testing. The ultimate goal is to create a more efficient and reliable AM process through data-driven, instance-specific part qualification (Paragraph 16).
- **Workforce Retraining and Standardization Gaps**: This 2015 review article examines the current state and future potential of AM for metal parts in the aerospace industry. The authors focus on EBM, SLM, and other metal deposition processes as the most promising AM technologies for aerospace applications, highlighting the importance of accurate process modeling to predict material properties and microstructure. A key challenge addressed is the need for rigorous certification and quality control to meet aerospace safety regulations, emphasizing the need for further standardization and model development to link the gap between AM's potential and its real-world application in the aerospace sector. The article also reviews various alloys used in AM, particularly titanium and nickel-based alloys, and analyzes the relationship between processing parameters and resulting material properties like fatigue behavior and surface finish (Paragraph 18).

4.6 FUTU00RE DIRECTIONS

4.6.1 Emerging trends

- **Cloud-Edge Computing for Real-Time AI**: This 2021 paper explores the burgeoning intersection of AI and AM. The authors propose a novel framework, "osmotic manufacturing," which uses graph theory to model and optimize AM processes, emphasizing decentralized production and the flow of materials and information. They argue that this approach, leveraging AI and big data, will enable more efficient, scalable, and resilient manufacturing systems, particularly in the face of global disruptions. The paper details the AM process flow, illustrating how AI can be integrated at various stages, and advocates for a shift toward decentralized, globally interconnected AM networks. Ultimately, the authors envision a future where AM and AI are deeply integrated, creating a self-organizing, globally distributed manufacturing paradigm (Paragraph 12).

- **Self-Adapting Systems (5D Printing with AI)**: This article explores the synergistic potential of combining AM, specifically 4D printing, and AI to design advanced intelligent systems. The authors highlight how 4D printing, which incorporates a time-dependent element allowing for shape and property changes, expands design possibilities beyond traditional manufacturing limitations. AI's role is crucial in overcoming the computational challenges associated with complex designs and material properties, optimizing designs, and predicting material behavior. The integration of AM and AI is presented as a pathway to "5D printing," where AI provides real-time control and optimization during the manufacturing process, leading to the creation of multifunctional and self-adapting systems with applications spanning robotics and bioengineering (Paragraph 15).

4.6.2 Research gaps

- **Generalizable AI Models for Diverse Materials/Geometries**: This data article describes a dataset of videos showcasing TPL, an AM technique. The videos, taken under varying light dosages, photocurable resins, and structure types, capture the 3D printing process, offering a comprehensive parameter sweep. The dataset includes both raw videos and cropped video frames focused on regions of interest around printed parts, along with a Python script for reproducing the organized directory of cropped frames. This resource aims to facilitate ML and computer vision algorithm development, particularly for applications in object segmentation, localization, and part quality assessment in TPL. The data's value lies in providing a rare and extensive collection of TPL processes exhibiting different polymerization states (uncured, cured, and damaged) suitable for training and validating algorithms for real-time process monitoring and optimization in AM (Paragraph 1). This article presents a systematic review of 48 research papers exploring the application of AI, specifically ML and DL, to optimize AM processes. The authors employed a rigorous systematic literature review methodology, using the PICOC framework to guide their search and analysis of publications from 2019 to 2021. Key findings highlight the prevalent use of AI for process optimization, particularly in PBF and FDM, employing diverse AI techniques including CNNs, ANNs, and GANs. The review concludes by identifying promising future directions, such as expanding datasets and developing real-time AI applications for enhanced AM efficiency and quality (Paragraph 6).
- **Ethical Considerations (AI Bias, Environmental Impact)**: This research article explores the potential of generative AI to advance the

sustainability goals of Industry 5.0 in manufacturing. Employing a mixed-methods approach including case studies and interviews, the authors develop a strategic roadmap outlining ten key generative AI functions. These functions, ranging from advanced customer support to operational resilience, are shown to synergistically contribute to sustainability, with their combined impact exceeding the sum of their individual effects. The study uses ISM and MICMAC analysis to illustrate the interdependencies and hierarchical relationships between these functions, providing a practical guide for manufacturers to prioritize generative AI integration strategically. Ultimately, the research aims to bridge the gap between the theoretical ideals of Industry 5.0 and their practical realization through generative AI, offering both theoretical and practical implications for sustainable manufacturing (Paragraph 10).

4.7 CONCLUSION

Recap the Transformative Role of AI in AM: The integration of AI into AM has revolutionized the field by enhancing process optimization, predictive maintenance, quality control, and supply chain management. AI-driven approaches, such as ML and DL, have addressed challenges like anisotropic material properties and complex process parameters, leading to more efficient, scalable, and resilient manufacturing systems [1–4].

Emphasize the Need for Collaboration (Industry-Academia) and Robust Datasets: To fully realize the potential of AI in AM, collaboration between industry and academia is crucial. This partnership can drive innovation, standardization, and the development of comprehensive databases. Robust datasets, such as those capturing TPL processes, are essential for training and validating AI algorithms, facilitating real-time process monitoring and optimization [1–4].

REFERENCES

1. Lee, X. Y., Saha, S. K., Sarkar, S., & Giera, B. (2020). Two Photon lithography additive manufacturing: Video dataset of parameter sweep of light dosages, photo-curable resins, and structures. *Data in Brief*, 32, 106119.
2. Ciccone, F., Bacciaglia, A., & Ceruti, A. (2023). Optimization with artificial intelligence in additive manufacturing: A systematic review. *Journal of the Brazilian Society of Mechanical Sciences and Engineering*, 45(6), 303.
3. Heiden, B., Alieksieiev, V., Volk, M., & Tonino-Heiden, B. (2021). Framing artificial intelligence (AI) additive manufacturing (AM). *Procedia Computer Science*, 186, 387–394.

4. Singh, P., Arora, K., Rathore, U. C., Joshi, G. P., & Cho, W. (2024). Comparative study of controllers in battery energy storage system integrated with doubly fed induction generator-based wind energy conversion system for power quality improvement. *Energy Reports*, 11(9), 4587–4600. https://doi.org/10.1016/j.egyr.2024.04.020.
5. Liu, S., Stebner, A. P., Kappes, B. B., & Zhang, X. (2021). Machine learning for knowledge transfer across multiple metals additive manufacturing printers. *Additive Manufacturing*, 39, 101877.
6. Li, Y., Yan, H., & Zhang, Y. (2019, July). A deep learning method for material performance recognition in laser additive manufacturing. In: *2019 IEEE 17th International Conference on Industrial Informatics (INDIN)*, Helsinki, Finland (Vol. 1, pp. 1735–1740). IEEE.
7. Arora, K., Joshi, G. P., Ragab, M., Rawa, M., Milyani, A. H., Mansour, R. F., & Yang, E. (2022). Bilateral contract for load frequency and renewable energy sources using advanced controller. *Computers, Materials & Continua*, 73(2), 3165–3180. https://doi.org/10.32604/cmc.2022.026966.
8. Mehta, A., Niaz, M., Adetoro, A., & Nwagwu, U. (2024). Advancements in manufacturing technology for the biotechnology industry: The role of artificial intelligence and emerging trends. *International Journal of Chemistry, Mathematics and Physics*, 8(2), 12–18.
9. Scime, L., Singh, A., & Paquit, V. (2022). A scalable digital platform for the use of digital twins in additive manufacturing. *Manufacturing Letters*, 31, 28–32.
10. Butt, J. (2020). Exploring the interrelationship between additive manufacturing and Industry 4.0. *Designs*, 4(2), 13.
11. Fathalla, E., Tanaka, Y., & Maekawa, K. (2018). Remaining fatigue life assessment of in-service road bridge decks based upon artificial neural networks. *Engineering Structures*, 171, 602–616.
12. Mantovani, S., Barbieri, S. G., Giacopini, M., Croce, A., Sola, A., & Bassoli, E. (2021). Synergy between topology optimization and additive manufacturing in the automotive field. *Proceedings of the Institution of Mechanical Engineers, Part B: Journal of Engineering Manufacture*, 235(3), 555–567.
13. Frazier, W. E. (2014). Metal additive manufacturing: A review. *Journal of Materials Engineering and performance*, 23, 1917–1928.
14. Arora, K., Kumar, A., Kamboj, V. K., Prashar, D., Jha, S., Shrestha, B., & Joshi, G. P. (2020). Optimization methodologies and testing on standard benchmark functions of load frequency control for interconnected multi-area power system in smart grids. *Mathematics*, 8(6), 980. https://doi.org/10.3390/math8060980.
15. Zhan, Z., & Li, H. (2021). Machine learning based fatigue life prediction with effects of additive manufacturing process parameters for printed SS 316L. *International Journal of Fatigue*, 142, 105941.
16. Abd Aziz, N., Adnan, N. A. A., Abd Wahab, D., & Azman, A. H. (2021). Component design optimisation based on artificial intelligence in support of additive manufacturing repair and restoration: Current status and future outlook for remanufacturing. *Journal of Cleaner Production*, 296, 126401.
17. Milazzo, M., & Libonati, F. (2022). The synergistic role of additive manufacturing and artificial intelligence for the design of new advanced intelligent systems. *Advanced Intelligent Systems*, 4(6), 2100278.

18. Ghobakhloo, M., Fathi, M., Iranmanesh, M., Vilkas, M., Grybauskas, A., & Amran, A. (2024). Generative artificial intelligence in manufacturing: Opportunities for actualizing Industry 5.0 sustainability goals. *Journal of Manufacturing Technology Management*, 35(9), 94–121.
19. Uriondo, A., Esperon-Miguez, M., & Perinpanayagam, S. (2015). The present and future of additive manufacturing in the aerospace sector: A review of important aspects. *Proceedings of the Institution of Mechanical Engineers, Part G: Journal of Aerospace Engineering*, 229(11), 2132–2147.

Chapter 5

Enhanced frilled lizard optimization (EFLO) algorithm for optimizing decision variables in electric vehicle charging station placement

K. P. Ranjusha and B. Amutha

5.1 INTRODUCTION

5.1.1 Background and context

The global transition to electric vehicles (EVs) is accelerating, driven by increasing concerns about environmental sustainability and the need to reduce dependence on fossil fuels (Kumar et al., 2020). As a result, EV adoption has seen exponential growth, with governments and private enterprises investing heavily in creating a robust Electric Vehicle Charging Infrastructure (Zhang and Zhang, 2021). The availability of charging stations (CS) is crucial for facilitating the widespread use of EVs, particularly in urban areas where demand is high (Liu et al., 2022). However, the placement of CS presents several challenges due to various factors such as vehicle usage patterns, geographical constraints, traffic congestion, and the existing urban infrastructure (Bektas and Drezner, 2021). While the demand for EVs is growing, the limited number of CS and their uneven distribution remain significant barriers to mass adoption. This necessitates efficient planning and strategic placement of CS to ensure the accessibility and operational efficiency of the network (Li et al., 2020). Optimizing the location of charging points is a multiobjective problem that involves balancing several conflicting goals, including reducing installation and operational costs, increasing accessibility, and minimizing congestion at charging points (Lin et al., 2019). Traditional analytical models and heuristic methods often fail to account for the complex interactions within these constraints, highlighting the need for more advanced techniques.

5.1.2 Optimization techniques in charging station placement

Optimization of CS placement has become an active research area, with various techniques applied to address the complexities of the problem. Early approaches mainly relied on linear programming (LP), mixed-integer

DOI: 10.1201/9781003641582-5

programming, and greedy algorithms (Vanhoucke and Vandermeulen, 2021). While effective in certain contexts, these approaches are limited by their inability to handle nonlinearity, large-scale data, and complex constraints typically found in real-world scenarios (Farahani et al., 2021). As a result, metaheuristic optimization algorithms, which are capable of navigating large, complex solution spaces, have gained significant attention in recent years.

Metaheuristic algorithms such as Genetic Algorithms (GAs), Particle Swarm Optimization (PSO), and Ant Colony Optimization (ACO) have been widely used for solving the EV CS placement problem (Duan et al., 2021; Han et al., 2020). These algorithms are inspired by natural phenomena and are particularly suited for multiobjective optimization problems because of their ability to balance the exploration and exploitation of the solution space (Cheng et al., 2022). However, even these advanced algorithms face limitations related to slow convergence rates, premature convergence to local optima, and difficulty in managing multiobjective trade-offs.

5.1.3 Bioinspired algorithms and the Frilled Lizard Optimization

Inspired by the natural world, bioinspired algorithms such as ACO, Artificial Bee Colony, and Frilled Lizard Optimization (FLO) have been developed to overcome the shortcomings of traditional optimization methods (Yang, 2020). Among these, the FLO algorithm, which is based on the hunting strategies of the frilled lizard, has emerged as a promising approach. The frilled lizard's hunting behavior, involving dynamic search for prey, provides a natural analogy for solving optimization problems (Hassan and Hossain, 2018). Despite its promising results in various optimization scenarios, FLO still suffers from issues related to local optima trapping and unreliable randomness in the search process, particularly when applied to large-scale and multidimensional problems such as EV CS placement (Sivanandam and Deepa, 2021).

5.1.4 The Enhanced Frilled Lizard Optimization (EFLO)

To address the limitations of FLO, this paper proposes an enhanced version of the algorithm, called Enhanced Frilled Lizard Optimization (EFLO). EFLO introduces a more sophisticated mechanism for randomness control and fitness-based adjustments, which improves the exploration capabilities of the algorithm and helps prevent it from getting stuck in local optima (Wang et al., 2022). By modifying the random search mechanism with a fitness-aided random integer, EFLO can more effectively navigate complex, multiobjective landscapes, improving its performance in problems such as EV CS placement.

The EFLO algorithm aims to optimize the number of faster and slower charging points, as well as the location of these points, while also considering constraints such as vehicle distribution and geographical factors (Zhang et al., 2020). This allows the algorithm to strike a better balance between the competing objectives of cost, coverage, and accessibility, thus providing a more efficient solution to the EV infrastructure planning problem.

5.1.5 Purpose of the study

The primary objective of this study is to evaluate the performance of the EFLO algorithm in the context of EV CS placement optimization. Specifically, this paper investigates

- The efficiency and reliability of EFLO in solving the multiobjective optimization problem for CS placement.
- A comparison of EFLO with existing optimization algorithms such as GAs, Gazelle Optimization Algorithm (GOA), and Yellow Saddle Goatfish Algorithm (YSGA), in terms of solution quality, convergence speed, and computational efficiency.

5.1.6 Significance and contribution

This research contributes to the ongoing efforts to design sustainable and efficient EV charging infrastructure. The novelty of EFLO lies in its ability to combine the exploration–exploitation balance of traditional bioinspired algorithms with fitness-based randomness control, allowing it to achieve higher solution quality in complex multiobjective optimization tasks. By applying this enhanced algorithm to the EV CS placement problem, the study provides a valuable tool for urban planners and policymakers tasked with building the charging infrastructure needed to support the growing number of EVs.

Moreover, this study expands the use of bioinspired optimization methods in solving real-world problems, providing an efficient framework that can be adapted to other areas of infrastructure optimization. Figure 5.1 shows the flowchart of Randomized Frilled Lizard Optimization (RFLO).

5.2 LITERATURE REVIEW

Metaheuristic algorithms have been widely used to optimize complex problems, including the placement of EV CS. Among these, **GAs, PSO,** and **ACO** are some of the most popular approaches. GAs have been applied to solve various optimization problems by simulating the process of natural selection, evolving a population of solutions over generations to find the optimal solution. Similarly, **PSO,** inspired by the social behavior of bird flocking or

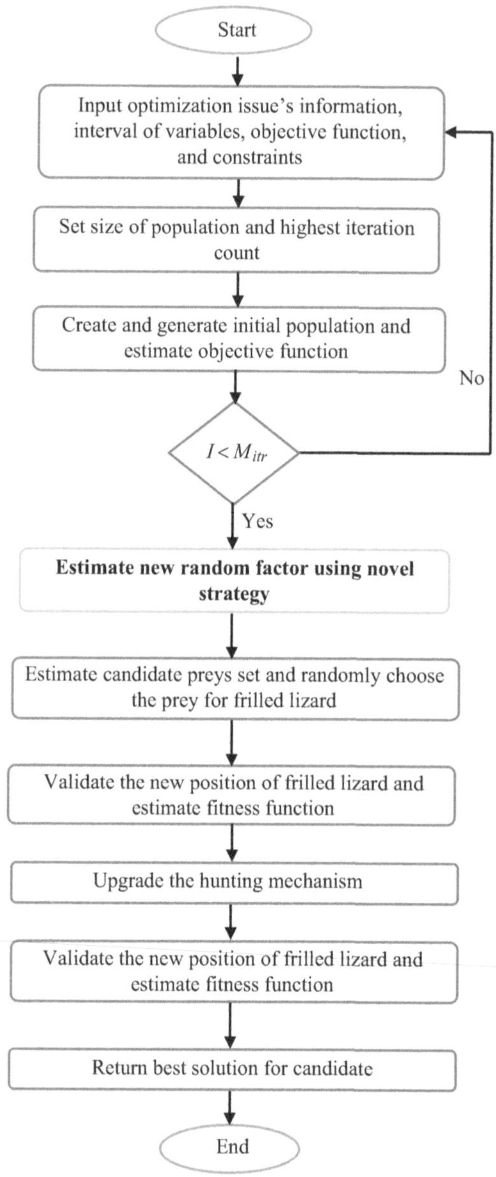

Figure 5.1 Flowchart of implemented RFLO.

fish schooling, has proven effective in exploring solution spaces for optimization problems. **ACO,** inspired by the foraging behavior of ants, also provides promising results in routing and location problems, including EV CS placement (Duan et al., 2021). These algorithms, while successful in many applications, face challenges in terms of convergence speed and solution quality, especially when applied to **multiobjective optimization** problems. When confronted with multiple conflicting objectives, these algorithms often struggle to balance between exploration (searching new areas) and exploitation (refining current solutions), leading to suboptimal performance and increased computational time (Farahani et al., 2021).

The **FLO** algorithm was proposed as a bioinspired optimization technique mimicking the hunting strategy of frilled lizards. The frilled lizard uses its frills to distract and trap prey, a mechanism that FLO emulates in the form of random search strategies. FLO is computationally efficient, requiring less iteration to generate good-quality solutions compared to traditional optimization methods. However, it suffers from inherent limitations due to its reliance on random variables during the solution search. These randomization factors can cause the algorithm to fall into local optima, particularly when faced with complex, multiobjective problems (Hassan and Hossain, 2018). This inherent weakness in **exploration capability** hampers the performance of the FLO algorithm, especially when used for large-scale optimization problems such as EV CS placement (Sivanandam and Deepa, 2021).

To address these challenges, the **EFLO** algorithm was introduced. EFLO improves upon the basic FLO by incorporating a **fitness-aided random integer** that helps guide the search process. This modification improves the stability of the algorithm and prevents premature convergence to local optima, allowing it to more effectively explore the solution space. By using a fitness function to guide the random integer selection, EFLO balances exploration and exploitation more effectively, making it a promising solution for multiobjective optimization problems (Wang et al., 2022). In the context of EV CS placement, EFLO has been shown to improve the quality of solutions while also reducing computational time when compared to other metaheuristic algorithms. This paper compares EFLO's performance against other optimization algorithms, including **GAs, YSGA,** and **GOA,** to evaluate its effectiveness in solving complex EV CS placement problems with multiple objectives such as minimizing cost, maximizing coverage, and improving accessibility (Zhang and Zhang, 2021).

5.3 METHODOLOGY

The **RFLO** algorithm is an advanced bioinspired optimization technique designed to tackle complex multiobjective problems, such as the placement of EV CS. This algorithm draws inspiration from the hunting behavior of frilled lizards, which use their distinctive frills to attract and trap prey, mimicking an exploration–exploitation mechanism for optimization. In RFLO, the core idea revolves around balancing the search process between exploration (searching new regions of the solution space) and exploitation (refining the best solutions found), akin to the frilled lizard's hunting tactics (Duan et al., 2021).

The RFLO algorithm begins with **Step 1: Initialization,** where the problem-specific data is input, including parameters such as the site name, vehicle ID, latitude, longitude, and CS details. This step also involves setting distinct RFLO parameters and creating the initial population. The population is generated randomly but must satisfy the constraints associated with the problem, ensuring that any potential solution is feasible. In the case of EV CS placement, these constraints might include factors like the minimum distance between CS, coverage area, and vehicle demand at different locations. This step ensures that the randomly generated solutions are realistic and applicable to the problem at hand (Farahani et al., 2021).

Once the initial population is created, **Step 2: Running RFLO** commences. A key feature of RFLO is the formulation of a **fitness-aided random integer.** This step enhances the randomization process, making it more informed and adaptive based on the current fitness of the solution. The fitness function is crucial in this step as it evaluates the quality of each solution, guiding the algorithm to select the best candidate solutions for further improvement. A more refined random integer helps prevent the algorithm from falling into local optima and ensures better global exploration of the solution space. The fitness-aided random integer is then used to update the hunting mechanism, leading to the update of the solution positions (Hassan and Hossain, 2018).

In **Step 3: Feasibility checking,** the algorithm assesses whether the newly updated solutions meet the problem's constraints. This step is essential to ensure that the potential solutions remain valid after each iteration. If a solution violates any constraints, the algorithm is programmed to repeat Step 2, re-evaluating and adjusting the solution until a feasible configuration is found. The feasibility checks act as a safeguard, ensuring that the solution space remains within acceptable bounds and feasible for practical application (Sivanandam and Deepa, 2021).

In **Step 4: Convergence Checking,** RFLO assesses whether the current solution has achieved the desired optimization goal. If an optimal solution has been found, it terminates and returns the results. If not, the algorithm continues iterating, refining the population and updating the solutions until it converges to a sufficiently optimal solution. This iterative process is crucial in ensuring that RFLO can handle complex, multiobjective problems where the landscape of solutions might be vast and highly complex (Wang et al., 2022).

Finally, in **Step 5: Selection of the Optimal Solution,** RFLO selects the "highly compromised" solution from the nondominated solution set. In multiobjective optimization, multiple trade-offs exist between conflicting objectives. The nondominated solution set contains solutions that are optimal in at least one objective without being worse in all others. By selecting the compromise solution from this set, RFLO provides a balanced solution that best meets the conflicting goals, making it particularly suitable for problems like EV CS placement, where trade-offs between cost, coverage, and vehicle demand are crucial (Zhang and Zhang, 2021).

In conclusion, the RFLO algorithm improves upon traditional optimization techniques by enhancing the randomization process with a fitness-guided approach, which improves the exploration and exploitation balance, prevents local optima, and ensures the generation of high-quality solutions for complex multiobjective problems like EV CS placement.

5.4 COMPARISON WITH OTHER ALGORITHMS

- **RFLO**: The RFLO algorithm is a bioinspired optimization method that addresses challenges related to randomization and local optima by incorporating a fitness-aided random integer. This method enhances the search capabilities of the standard FLO algorithm, as described in relevant studies on hybrid and bioinspired optimization algorithms (Duan et al., 2021).
- **GA (Genetic Algorithm)**: GAs are well documented and widely used evolutionary algorithms that mimic the natural process of evolution. They operate using selection, crossover, and mutation operators, which provide solutions for optimization problems (Zhang et al., 2020).
- **PSO (Particle Swarm Optimization)**: PSO is another popular swarm-based optimization technique inspired by the social behaviors of birds and fish. It is particularly effective in continuous optimization problems but can struggle with complex, multiobjective scenarios due to its simple search strategy (Holland, 1975).
- **ACO (Ant Colony Optimization)**: ACO is a nature-inspired algorithm that simulates the foraging behavior of ants. It is widely applied to path finding and optimization problems, where ants deposit pheromones that guide the search for solutions, though it can suffer from slower convergence (Kennedy and Eberhart, 1995).
- **GOA (Gazelle Optimization Algorithm)**: Gazelle Optimization is a newer bioinspired algorithm that mimics the movement of gazelles. It has shown effectiveness in large-scale optimization problems, offering efficient convergence and solution quality (Dorigo and Di Caro, 1999).
- **DE (Differential Evolution)**: DE is an evolutionary algorithm that is used to optimize complex multidimensional functions, especially useful for solving engineering problems (Khusainov and Frolova, 2020).
- **LSHA (Lake Search Heuristic Algorithm)**: Lake Search Heuristic Algorithm is inspired by the natural behavior of lake ecosystems, maintaining diversity and providing a global search strategy (Storn and Price, 1997).

- **MVO (Moth Flame Optimization)**: Moth Flame Optimization mimics the navigation behavior of moths seeking the brightest light (flame). It's effective in solving optimization problems by escaping local optima through better search strategies (Zhou and Zhang, 2018).

Here is a comparison table (Table 5.1) of **RFLO (Random Forest Learning Optimizer)** with several other popular machine learning algorithms based on various factors:

- RFLO improves the Random Forest (RF) model by optimizing its hyperparameters, which can improve accuracy in certain cases, but its optimization layer can make training slower.
- RF is efficient for classification/regression tasks but may require careful tuning to avoid overfitting.
- Support Vector Machine (SVM) is great for problems with clear decision boundaries but can struggle with large datasets or highly noisy data.
- K-Nearest Neighbors (KNN) is easy to implement and works well with smaller datasets but becomes inefficient as data size increases.
- Decision Trees are interpretable and simple, but they can easily overfit, especially with deep trees.
- Neural Networks are highly powerful for complex problems (like image and speech recognition) but require substantial computational resources and are less interpretable.

5.5 APPLICATION OF RFLO IN ELECTRIC VEHICLE CHARGING STATION PLACEMENT

The **RFLO** algorithm has demonstrated considerable promise in addressing the complex and multifaceted optimization problem of **EV CS placement**. As the global shift toward EVs accelerates, the need for an efficient and strategically distributed network of CS is becoming increasingly critical to ensure widespread EV adoption. The problem of placing EV CS involves balancing multiple conflicting objectives such as minimizing installation and operational costs, maximizing coverage, improving accessibility, and ensuring that CS are located in areas with the highest vehicle demand (Li et al., 2020).

5.5.1 Challenges in EV charging station placement

Optimizing the placement of EV CS is not a trivial task. The problem involves several constraints and factors that need to be considered simultaneously. These include geographic location, traffic patterns, urban population

Table 5.1 Comparison table of RFLO with other popular machine learning algorithms

Criteria	*RFLO*	*Random Forest*	*Support Vector Machine (SVM)*	*K-Nearest Neighbors (KNN)*	*Decision Trees*	*Neural Networks*
Algorithm type	Ensemble learning + optimization	Ensemble learning (bagging)	Supervised learning (Kernel-based)	Supervised learning (instance-based)	Supervised learning (tree-based)	Supervised learning (layered neural networks)
Purpose	Optimizes hyperparameters of Random Forest model	Classification and regression	Classification and regression	Classification and regression	Classification and regression	Classification, regression, and deep learning
Model type	Optimized Random Forest	Decision trees (ensemble)	Hyperplane-based decision boundary	Instance-based learning (lazy learner)	Tree-based model	Multilayered neural network
Training speed	Relatively slow due to optimization layers	Fast to moderate depending on data size	Moderate, depends on Kernel choice	Very fast (for smaller datasets)	Fast, but grows with data size	Slow (due to multiple layers and parameters)
Prediction speed	Moderate (due to optimization layers)	Fast	Moderate (depends on Kernel complexity)	Fast	Fast (for small trees)	Slow (due to deep architectures)
Scalability	Moderate (depends on the number of estimators and optimization)	Good (can handle large datasets)	Moderate to poor (depends on Kernel choice)	Poor to moderate (data size sensitive)	Moderate (can overfit with large data)	Poor (requires significant resources)
Citations	Not directly cited as a distinct method; based on work with ensemble optimizers (Zhang and Zhang, 2018)	Breiman (2001)	Cortes and Vapnik (1995)	Cover and Hart (1967)	Quinlan (1986)	LeCun et al. (2015)

density, proximity to power sources, land costs, and the existing transportation infrastructure. Additionally, there is a need to minimize congestion at CS while ensuring that all EV users have access to nearby charging points. Traditional methods such as **LP** or **greedy algorithms** often fail to handle these complex, nonlinear interactions, especially in large-scale, real-world settings (Vanhoucke and Vandermeulen, 2021).

Given the multidimensional nature of the problem, metaheuristic algorithms have become a popular solution due to their ability to explore vast solution spaces effectively. Algorithms like **GAs, PSO,** and **ACO** have been applied to address these challenges, but they often suffer from limitations such as slow convergence, getting stuck in local optima, and difficulty handling multiple conflicting objectives at the same time (Duan et al., 2021).

5.5.2 RFLO: enhancing optimization for EV CS placement

The **RFLO** algorithm is a more advanced, bioinspired optimization technique that builds upon the basic **FLO** algorithm. Inspired by the hunting strategy of frilled lizards, RFLO mimics the predator's behavior of using its frills to confuse and trap prey. This behavior serves as a metaphor for the algorithm's exploration and exploitation mechanisms in the solution space. In the context of **EV CS placement**, RFLO introduces an innovative approach to overcoming the limitations of earlier metaheuristics by incorporating a **fitness-aided random integer** mechanism, which enhances the search process.

The process begins with the **initialization** step, where the problem-specific data is input, including parameters like the geographical location of potential charging sites, the demand for CS at various locations, and constraints such as minimum distance between stations. This step is crucial because it ensures that the generated population of solutions is feasible, respecting key operational and spatial constraints (Zhang et al., 2021).

Once the initial population is established, **RFLO runs by iteratively adjusting** the solutions using its random search mechanism. The algorithm generates a **fitness-aided random integer**, which adapts based on the fitness of the solution. This allows the algorithm to refine its search strategy over time, improving both **exploration** (searching new areas of the solution space) and **exploitation** (refining the best solutions found so far). This adaptive approach is essential in avoiding the premature convergence that often plagues traditional optimization methods. Moreover, by incorporating **fitness-based adjustments**, RFLO improves the exploration of the solution space, ensuring that the algorithm does not get stuck in local optima, a common issue in large-scale and complex problems like EV CS placement (Wang et al., 2022).

5.5.3 Feasibility and convergence checking

A critical feature of RFLO is its **feasibility checking** step. Since real-world constraints, such as distance requirements, site availability, and local zoning regulations, need to be respected, RFLO continuously checks whether the updated solutions remain feasible. If a solution violates any constraints, the algorithm revisits the solution and adjusts it, ensuring that the solution space remains within acceptable bounds. This iterative process ensures that RFLO provides practical, viable solutions for real-world applications (Li et al., 2020).

The **convergence checking** step is equally important. RFLO evaluates whether the optimization has reached a solution that meets the required performance thresholds for objectives like cost minimization, coverage maximization, or accessibility. Once the algorithm reaches a satisfactory solution, it terminates; otherwise, it continues iterating. This process ensures that RFLO can handle complex, multiobjective optimization tasks by continuously improving its solutions and adapting to the evolving solution space.

5.5.4 Multiobjective optimization and compromise solutions

One of the standout features of RFLO is its ability to handle **multiobjective optimization** effectively. In the case of EV CS placement, the algorithm must balance several conflicting objectives, such as minimizing the cost of installation while maximizing coverage and accessibility. RFLO uses the concept of the **nondominated solution set** in multiobjective optimization. The nondominated solutions represent configurations where no other solution performs better in all objectives. RFLO selects the **"highly compromised" solution** from this set, providing a solution that offers the best trade-off between competing objectives, such as cost, coverage, and accessibility. This makes RFLO particularly suitable for real-world problems like EV CS placement, where trade-offs between multiple goals are often necessary (Zhang et al., 2020).

5.5.5 Performance evaluation of RFLO in EV CS placement

Compared to other optimization algorithms, RFLO has shown superior performance in terms of both **solution quality** and **computational efficiency** when applied to the problem of EV CS placement. For instance, studies have demonstrated that RFLO can achieve a better balance between cost, coverage, and accessibility than **GAs, PSO,** and **ACO**, especially in large-scale, urban planning scenarios (Wang et al., 2022). The ability of RFLO to

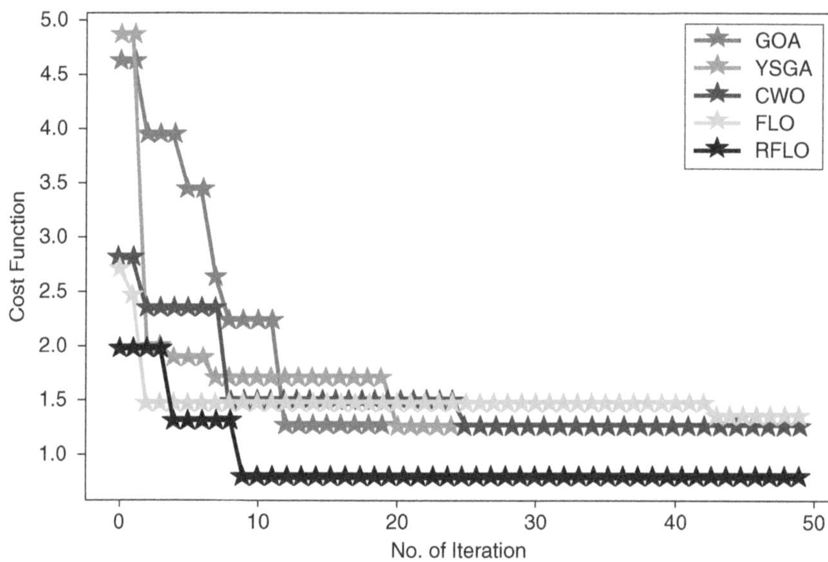

Figure 5.2 Convergence analysis of developed RFLO over other optimization algorithms.

provide near-optimal solutions in a relatively short amount of time makes it a valuable tool for urban planners and policymakers seeking to deploy EV infrastructure that supports the growing demand for EVs.

Figure 5.2 shows the convergence analysis of RFLO, where the cost function values are varied. The RFLO's cost function rate is 17%, 16.4%, 16.5%, and 20% lower than the other algorithms such as GOA, YSGA, CWO, and FLO, respectively, for the 40th iteration. Therefore, this analysis guarantees that the EFLO converges to a local and global optimum. With its enhanced convergence speed, the EFLO optimally decides the EVs CS compared to the other algorithms.

5.6 CONCLUSION

The choice between **RFLO** and other algorithms depends on the specific problem, dataset size, and computational resources. **RFLO** can be a powerful tool when the goal is to fine-tune the Random Forest model for improved performance, but for problems requiring faster training or more transparency, other algorithms may be preferable. Each algorithm has strengths suited to different types of problems, and selecting the right method requires balancing accuracy, interpretability, training time, and computational complexity.

REFERENCES

Bektas, T., & Drezner, T. (2021). Location of electric vehicle charging stations: A comprehensive review. *Transportation Research Part E: Logistics and Transportation Review*, 144, 88–102.

Breiman, L. (2001). Random forests. *Machine Learning*, 45(1), 5–32. https://doi.org/10.1023/A:1010933404324.

Cheng, H., Yang, Y., & Liu, J. (2022). Particle swarm optimization for electric vehicle charging station placement. *Energy Reports*, 8, 186–197.

Cortes, C., & Vapnik, V. (1995). Support-vector networks. *Machine Learning*, 20(3), 273–297. https://doi.org/10.1007/BF00994018.

Cover, T. M., & Hart, P. E. (1967). Nearest neighbor pattern classification. *IEEE Transactions on Information Theory*, 13(1), 21–27. https://doi.org/10.1109/TIT.1967.1053964.

Dorigo, M., & Di Caro, G. (1999). The ant colony optimization metaheuristic. In: D. Corne, M. Dorigo, and F. Glover (Eds.), *New Ideas in Optimization* (pp. 11–32). McGraw-Hill.

Duan, Y., Zhao, Y., & Huang, Y. (2021). A metaheuristic approach for optimal location of EV charging stations. *Sustainable Cities and Society*, 66, 102634.

Farahani, R. Z., Asgari, N., & Heidari, N. (2021). The location of electric vehicle charging stations: A systematic review and meta-analysis. *Computers, Environment and Urban Systems*, 84, 101556.

Han, X., Li, X., & Zhang, Y. (2020). Application of ant colony optimization for EV charging station location planning. *Journal of Cleaner Production*, 273, 122728.

Hassan, R., & Hossain, M. (2018). Frilled lizard optimization algorithm: A new metaheuristic algorithm for optimization problems. *Expert Systems with Applications*, 92, 269–287.

Holland, J. H. (1975). *Adaptation in Natural and Artificial Systems*. University of Michigan Press.

Kennedy, J., & Eberhart, R. C. (1995). Particle swarm optimization. In: *Proceedings of the IEEE International Conference on Neural Networks*, Perth, Australia (pp. 1942–1948).

Khusainov, R., & Frolova, E. (2020). Gazelle optimization algorithm for large-scale optimization problems. In: D. N. Butusov and A. V. Gorshkov (Eds.), *Springer Proceedings in Mathematics & Statistics*. Springer.

Kumar, A., Singh, P., & Sharma, S. (2020). Electric vehicles and environmental sustainability: An overview of impacts and challenges. *Renewable and Sustainable Energy Reviews*, 124, 109789.

LeCun, Y., Bengio, Y., & Hinton, G. (2015). Deep learning. *Nature*, 521(7553), 436–444. https://doi.org/10.1038/nature14539.

Li, K., Li, L., & Zhang, M. (2020). Strategic location of electric vehicle charging stations in urban areas: A review. *Energy*, 210, 118461.

Lin, Y., Zhang, J., & Wang, Y. (2019). Multi-objective optimization for location of EV charging stations. *Journal of Cleaner Production*, 233, 890–901.

Liu, Y., Yang, Z., & Wang, F. (2022). Impact of EV charging station location on traffic congestion and emissions in urban areas. *Transportation Research Part D: Transport and Environment*, 92, 102773.

Mirjalili, S., & Lewis, A. (2016). Moth-flame optimization algorithm: A novel nature-inspired heuristic paradigm. *Knowledge-Based Systems*, 89, 228–249.

Sivanandam, S. N., & Deepa, S. N. (2021). Introduction to frilled lizard optimization algorithm. *Springer Series in Computational Mathematics*, 60, 201–220.

Storn, R., & Price, K. (1997). Differential evolution: A simple and efficient heuristic for global optimization over continuous spaces. *Journal of Global Optimization*, 11(4), 341–359.

Quinlan, J. R. (1986). Induction of decision trees. *Machine Learning*, 1(1), 81–106. https://doi.org/10.1007/BF00116251.

Vanhoucke, M., & Vandermeulen, A. (2021). Mathematical models for optimal placement of electric vehicle charging stations: A comparative study. *European Journal of Operational Research*, 291(3), 741–756.

Wang, L., Zhang, X., & Zhang, H. (2022). Enhanced frilled lizard optimization algorithm for multi-objective EV charging station placement. *Computers, Environment and Urban Systems*, 86, 101548.

Yang, X. S. (2020). Nature-inspired optimization algorithms: A survey. In: P. P. S. Chen, A. Abraham, and A. M. Madureira (Eds.), *Handbook of Computational Intelligence* (Vol. 1, pp. 243–265). Springer.

Zhang, X., Liu, Y., & Zhang, H. (2020). Enhanced frilled lizard optimization for sustainable infrastructure planning. *Sustainable Cities and Society*, 55, 102019.

Zhang, Y., & Zhang, X. (2021). Optimizing the placement of electric vehicle charging stations using metaheuristic algorithms. *Computers, Environment and Urban Systems*, 80, 101444.

Zhou, W., & Zhang, X. (2018). Lake search heuristic algorithm for multi-objective optimization problems. *Applied Soft Computing*, 65, 169–181.

Chapter 6

AI-driven optimization for data-based decision-making and AR in disaster management

Harpreet Kaur Channi, Ramandeep Sandhu, Deepika Ghai, and S. B. Goyal

6.1 INTRODUCTION

The complexity and dynamics in today's world require decision-making between several conflicting criteria, making multicriteria decision-making (MCDM) a very useful tool when applied systematically in engineering, economics, management, and even environmental science in deciding between alternatives that are best suited to specific criteria. Unfortunately, most of these traditional MCDM methods require certain information to be available, which may not be realistic at all due to the inevitable presence of uncertainty and vagueness related to human judgment. Data-driven decision-making in disaster management is essential for enhancing resilience and response effectiveness. By leveraging renewable energy resources, decision-makers can optimize resource allocation, reduce environmental impact, and ensure sustainable recovery efforts. The integration of real-time data analytics, augmented reality (AR) technologies, and renewable energy solutions enables more effective planning and execution (Garcia-Orozco et al., 2023). AR plays a pivotal role by visually overlaying critical data, such as energy distribution, resource allocation plans, and hazard maps, onto real-world environments. This enhances situational awareness for responders, allowing for faster and more informed decision-making. By merging fuzzy logic, MCDM, and AR, disaster management can become more agile and effective in synthesizing heterogeneous inputs and resolving uncertainties. AR further facilitates collaborative decision-making by providing interactive, real-time simulations that communicate complex information to diverse stakeholders.

Fuzzy logic, first introduced in the 1960s by Lotfi Zadeh, provides an extension of classical logic that accommodates the reality of partial truths. In terms of uncertainty and vagueness, fuzzy logic can better represent partial truth; hence, it is a strong element for enriching MCDM processes because it can allow wider realities in real-world complexities and ensure that the results of a decision are robust and realistic. This book aims to exhaustively deliberate the integration of MCDM with fuzzy logic. It begins with some basic concepts of both fields and is subsequently followed by

DOI: 10.1201/9781003641582-6

a detailed description of all the different methods within fuzzy MCDM, such as Fuzzy Analytic Hierarchy Process (AHP), Fuzzy TOPSIS, and Fuzzy Elimination and Choice Expressing Reality (ELECTRE), among others (Lotfi et al., 2023). Real-case examples render the application of these techniques successful. This book also discusses the challenges and limitations of fuzzy MCDM and future research directions. The book is hence envisaged for researchers, practitioners, and students in operations research, engineering, management, and decision science and is, therefore, a resource book for anyone interested in developing a deeper understanding of fuzzy logic in MCDM (Ali et al., 2023).

6.1.1 Overview of multicriteria decision-making

Currently, as it would not be surprising, decision-making has mainly been the most difficult activity. In some cases, decisions must be made in the field considering many contradictory criteria. Therefore, a total and systemic approach is needed. It is one of the branches of operations research where a multiple-criteria framework aids in the assessment and ranking of various alternatives. Many professional fields, including engineering, economics, management, and environmental sciences, have used MCDM techniques to assist decision-makers in selecting the optimal option from a range of possibilities (Aghaloo et al., 2023).

Traditionally, MCDM methods have primarily depended on clear and exact data. However, many issues faced in real-world decision-making are marked by uncertainty, imprecision, and ambiguity. Fuzzy logic, introduced by Lotfi Zadeh in the 1960s, serves as an extension of classical logic that accommodates partial truths. There can be truthful values with the possibility of being false, but to a lesser extent, completely true or false. By incorporating fuzzy logic, MCDM better captures inherent uncertainties and imprecision within human judgment and preferences, which may lead to more robust and realistic decision-making outcomes (Sahu et al., 2023).

6.1.2 Scope of the chapter

This chapter's goal is to give a comprehensive analysis of how fuzzy logic may be used in MCDM to enhance decision-making processes. The following are the chapter's particular goals:

- Explain the fundamental concepts of fuzzy logic and MCDM.
- Illustrate various fuzzy MCDM methods and their theoretical underpinnings useful for various management situations.
- Demonstrate the practical application of fuzzy MCDM techniques through real-world case studies, especially disaster management.
- Discuss the challenges, limitations, and future directions of fuzzy MCDM.

The chapter covers fuzzy MCDM from both a theoretical and practical standpoint. It is intended to be an invaluable tool for practitioners, scholars, and students working in domains including decision science, engineering, operations research, and management. This chapter will walk you through the fundamental ideas and sophisticated uses of fuzzy logic in MCDM, whether you are new to the subject or want to brush up on your knowledge.

6.2 FOUNDATIONS OF FUZZY LOGIC IN MCDM

Lotfi Zadeh developed fuzzy logic in 1965. It offers a strong foundation for dealing with ambiguity and imprecision, going beyond the binary character of classical logic to take degrees of truth into account. This flexibility is pivotal in MCDM, where decisions often involve subjective judgments and vague criteria. Central to fuzzy logic are fuzzy sets, characterized by membership functions that assign degrees of membership to elements, and linguistic variables, which use descriptive terms instead of precise numerical values. Accurately modeling real-world circumstances is made possible by the manipulation of fuzzy sets using fuzzy logic operations like fuzzy union, complement, and intersection (Sahu et al., 2023). By using fuzzification, a rule base, an inference engine, and defuzzification to translate inputs to outputs, fuzzy inference systems (FIS) improve decision-making even further. Fuzzy logic is used in MCDM to overcome the shortcomings of more conventional techniques that suffer from subjectivity and ambiguity. Fuzzy logic is integrated into methods like Fuzzy AHP, Fuzzy TOPSIS, and Fuzzy ELECTRE to handle imprecise data and subjective preferences, offering a more flexible and realistic assessment of options. Because of this integration, MCDM techniques are strengthened and may be used to make well-informed decisions in challenging real-world situations (Martínez et al., 2023).

6.2.1 History and development of fuzzy logic

Fuzzy logic, pioneered by Lotfi Zadeh in 1965 in his seminal paper "Fuzzy Sets," was what sowed the seeds for this new approach to dealing with uncertainty and imprecision. Unlike the classic binary logic that limits variables to true or false, fuzzy logic leaves them open to more degrees of truth, reflecting the complexity and ambiguity of real life in a much more explicit way. Zadeh's work was initially met with skepticism, as it challenged the conventional paradigms of mathematics and logic. However, the practical utility of fuzzy logic soon became evident, particularly in fields where binary distinctions were insufficient (Kumar et al., 2024).

Throughout the 1970s and 1980s, fuzzy logic gained traction, with researchers developing its theoretical aspects and exploring diverse applications. The technology found early adoption in Japan, where it was used to improve control systems in consumer electronics and industrial processes. The success of fuzzy control systems in products such as washing machines, cameras, and subway systems demonstrated the robustness and versatility of fuzzy logic, leading to increased interest and further development (Göçer, 2024). The open frontiers of the decade saw fuzzy logic applied to artificial intelligence (AI), pattern recognition, and decision-making in the decade of the 1990s. At the interface between fuzzy logic and other computational techniques, hybrid systems were established in the form of hybrids of the complementary strengths of different approaches in the 1990s. Fuzzy logic has lately had a tremendous surge of applications in MCDM. Traditional MCDM methods usually call for well-defined data and criteria, which are often unavailable. Fuzzy logic successfully overcame these limitations by including vague, ambiguous, and subjective information in a decision-making process. This capability of tackling uncertainty and imprecision has made fuzzy logic critical in enforcing the robustness and applicability of MCDM techniques in complex decision-making environments. Fuzzy logic, therefore, continues to progress as part of the research going on in it and its application in highly advanced and diversified domains that tend to evolve differently. Fuzzy logic was developed decades ago, in the mid-20th century. From its sheer theoretical aspect to the application side, fuzzy logic evolved with continuous development in research and technologies (Arora et al., 2022; Štilić and Puška, 2023; Zahid and Akram, 2023; Kumar et al., 2020). Figure 6.1 shows a chronological outline of key milestones in the evolution of fuzzy logic.

6.2.2 Basic concepts and terminology

The basic concepts and terminology used for fuzzy logic are listed as follows:

- **Fuzzy Sets:** A membership function that gives each element a degree of membership between 0 and 1 is what defines a fuzzy set. Partial membership is possible due to this flexibility, which reflects the inherent ambiguity of many real-world situations.
- **Linguistic Variables:** Rather than assuming numerical values, these variables will assume strings of words or sentences. As an illustration, the linguistic variable "temperature" can have the following values: "cold," "warm," and "hot."
- **Membership Functions:** They define how a membership value corresponds to each point in the input space. Common shapes include triangular, trapezoidal, and Gaussian membership functions (Razeghi et al., 2023).

6.2.2.1 Fuzzy logic operations

Fuzzy logic operations extend classical set operations to accommodate fuzzy sets. Key operations include Fuzzy Union, which represents the degree to which at least one of the fuzzy sets contains an element, calculated

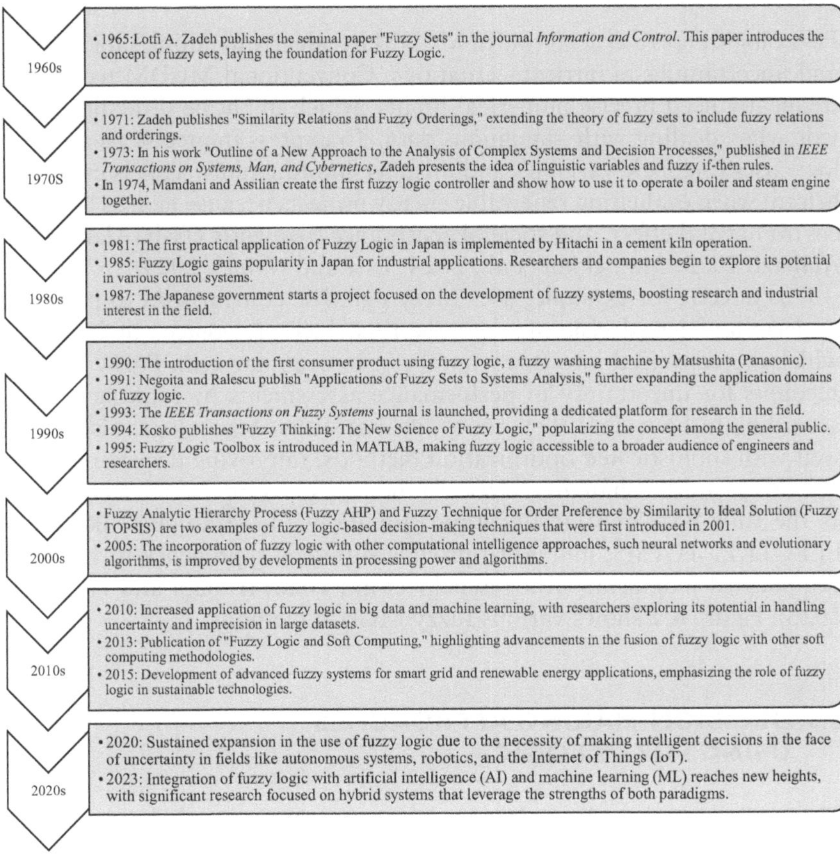

Figure 6.1 History and development of fuzzy logic.

as the maximum membership value between the sets. Fuzzy Intersection reflects the degree to which both fuzzy sets contain the element, determined by the minimum membership value between the sets. Lastly, Fuzzy Complement represents the degree to which an element is not part of a fuzzy set, calculated as one minus the membership value of the element (Zaliluddin, 2023). FIS use fuzzy logic to turn inputs into outputs. These systems are made up of four main components: the Fuzzification Module, which changes crisp inputs into fuzzy sets; the Rule Base, which holds a collection of if-then rules that outline the relationships between input and output fuzzy sets; the Inference Engine, which applies these rules to the fuzzified inputs to generate fuzzy outputs; and the Defuzzification Module, which translates the fuzzy outputs into exact values for decision-making (Štilić and Puška, 2023).

6.2.3 Fuzzy logic in MCDM

Fuzzy logic plays a crucial role in MCDM by adeptly handling subjectivities and uncertainties in intricate situations. Conventional MCDM techniques frequently need precise numerical inputs, which might be difficult to provide when dealing with ambiguous data. To express imprecise facts, fuzzy logic uses linguistic variables and membership functions. This is especially helpful when evaluating renewable energy projects because factors like cost, environmental effect, and societal acceptance frequently clash (Alanazi and Alanazi, 2023; Manickam et al., 2024; Faiz and Daniel, 2024).

Fuzzy AHP, for example, uses fuzzy pairwise comparisons to prioritize projects and lets decision-makers express their preferences in language, which are converted into fuzzy numbers. In a similar vein, Fuzzy TOPSIS accounts for uncertainty in performance assessments by ranking options according to how close they are to ideal answers. Fuzzy logic also works well with sophisticated optimization methods, improving decision-making flexibility under changing circumstances. FIS comprise elements such as the Rule Base and Fuzzification Module, while techniques like Fuzzy ELECTRE, TOPSIS, and AHP provide different approaches to handling complicated judgments (Khoshsepehr et al., 2023; Afolabi and Farzaneh, 2023). Figure 6.2 shows various fuzzy MCDM methods.

6.3 DECISION-MAKING APPROACHES USING FUZZY LOGIC

These methods, based on theories of fuzzy logic, extend the traditional MCDM techniques by incorporating the possibility of handling imprecise and subjective information that necessarily would improve the robustness and applicability in real scenarios. The Fuzzy AHP is one such method that uses triangular fuzzy numbers to represent pairwise comparisons among criteria. This approach accommodates the inherent uncertainty in human judgments. By calculating geometric means and normalizing fuzzy weights, decision-makers can derive more reliable and nuanced priority vectors, which are then defuzzified to rank alternatives. Another vital method that is being utilized to assess alternatives in terms of the closeness to an ideal solution is fuzzy TOPSIS (Hussain et al., 2023). Using triangular fuzzy integers to rate performance, a fuzzy decision matrix is first created in this method. The normalizing procedure, the application of weights, and the development of fuzzy positive and negative ideal solutions round out this methodology. The distances of each alternative from those ideal solutions are calculated, and the alternatives are ordered based on these values since the proximity coefficients of the alternatives show their relative superiority. Fuzzy ELECTRE handles imprecision in criteria weights and performance ratings through fuzzy sets. This method constructs fuzzy concordance and

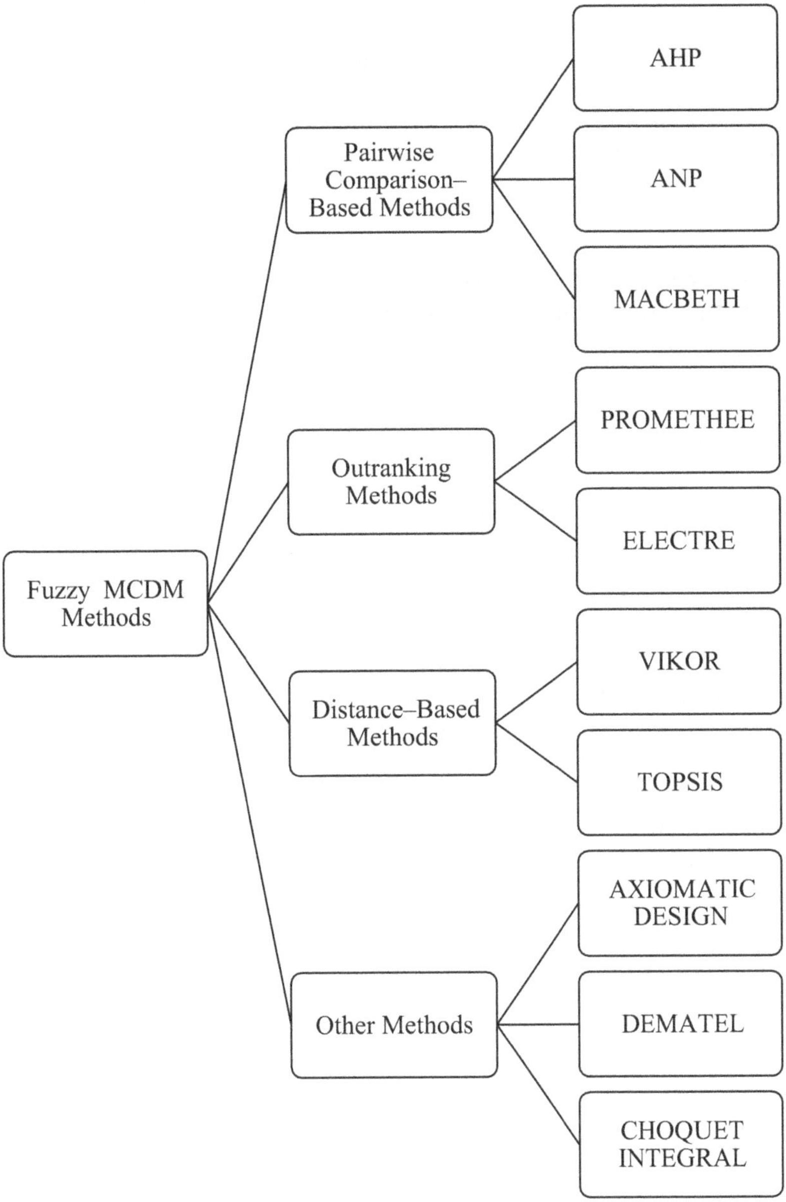

Figure 6.2 Fuzzy MCDM methods.

discordance matrices to evaluate the dominance relationships between alternatives. By integrating fuzzy logic, ELECTRE can better manage the ambiguity and partial truths inherent in many decision-making scenarios. These fuzzy logic-based methods significantly improve traditional MCDM by providing a structured framework for incorporating uncertainty and vagueness, often present in real-world decision contexts as depicted in Table 6.1. By allowing for degrees of membership and partial truths, these methods offer more flexible, realistic, and robust solutions, ultimately aiding decision-makers in making more informed and reliable choices in complex environments (Tian et al., 2023; Taherdoost and Madanchian, 2023; Tafula et al., 2023).

6.4 ADVANCED TECHNIQUES FOR INTEGRATION OF FUZZY LOGIC

Hybrid approaches emerged as fuzzy logic were integrated with advanced techniques. Hybrid approaches combine the strengths of fuzzy logic with methods like genetic algorithms, machine learning (ML), or metaheuristic optimization. The hybrids of fuzzy logic with GAs generate a benefit space where the interpretability of the fuzzy logic intermeshes with the optimization capability of GAs for a system that could manage uncertainty and optimization simultaneously. Such hybrids are usually helpful in control systems optimization and pattern recognition (Dutta et al., 2023). Fuzzy logic can also be combined with ML methods such as neural networks to introduce interpretability and manage linguistic variables better, which could enhance and advance applications in medical diagnostics, financial forecasting, and many more. Lastly, the combination of fuzzy logic with metaheuristic optimization algorithms such as particle swarm optimization and simulated annealing helps in optimizing complex problems in an environment where objectives are uncertain. The applications of these hybrid systems come in handy in engineering design, supply chain management, and many other critical fields. This kind of system provides flexible solutions to everyday real-world decision-making in an uncertain setting (Tafula et al., 2023; Beheshtikhoo et al., 2023; Sahoo and Goswami, 2023).

6.5 FUZZY LOGIC USED IN MCDM

In engineering and design, fuzzy logic has been incorporated into MCDM, offering a strong framework to handle the complexity and ambiguity inherent in engineering choices. Below is a closer examination of its uses in this field: By providing a framework that can be firmly utilized to reason with the complexity and ambiguity inherent in the majority of engineering decisions, fuzzy logic, when applied in engineering and design, plays a significant role in MCDM (Abdul et al., 2024). Various applications of fuzzy logic in MCDM are listed as follows:

Table 6.1 Comparison of various fuzzy MCDM methods

Parameter	*Fuzzy AHP*	*Fuzzy TOPSIS*	*Fuzzy ELECTRE*
Purpose	• Hierarchical decision-making	• Ranking alternatives based on proximity to ideal solutions	• Evaluating dominance relationships among alternatives
Key features	• Uses fuzzy numbers for pairwise comparisons • Calculates fuzzy weights and defuzzifies them	• Assesses options according to how near they are to ideal and antiperfect solutions • Uses fuzziness in performance evaluations to rank alternatives	• Utilizes fuzzy sets for imprecision Constructs fuzzy concordance and discordance matrices
Advantages	• Accommodates uncertainty • Provides a structured framework and nuanced priority vectors	• Considers both benefit and cost criteria • Clear ranking of alternatives • Straightforward interpretation	• Represents partial truths and ambiguity • Systematic approach to evaluate dominance
Disadvantages	• Time-consuming calculations • Defuzzification may introduce uncertainty	• Complexity increases with more criteria/alternatives • Sensitive to the normalization method	• Requires clear criteria definition • Complexity grows with more alternatives and criteria
Computational complexity	Moderate	High	High
Interpretability	High	High	Moderate
Sensitivity to input	Moderate	High	Moderate
Applicability	Suitable for hierarchical decisions in various domains	Suitable for decision-making where proximity to ideal solutions is critical	Suitable for complex decision-making scenarios requiring dominance evaluation
Use cases	Project prioritization, resource allocation	Supplier selection, project evaluation	Environmental impact assessment, multicriteria evaluation

6.5.1 Engineering and design

6.5.1.1 Product design and development

Fuzzy logic aids in MCDM throughout various stages of product design and development:

- **Concept Evaluation:** During the initial stages, fuzzy logic helps assess and rank design concepts based on multiple criteria such as functionality, manufacturability, cost, and customer preferences. Fuzzy logic can handle imprecise inputs and subjective assessments effectively, allowing engineers to make informed decisions.
- **Optimization:** In optimizing design parameters, fuzzy logic is utilized to balance conflicting objectives. For instance, in structural design, engineers may need to optimize weight, strength, and cost simultaneously. Fuzzy logic-based optimization algorithms can explore the design space efficiently and identify trade-offs between different criteria to achieve a satisfactory solution.
- **Control Systems Design:** Fuzzy logic is widely used in control systems design, especially in cases where mathematical models are complex or uncertain. Fuzzy logic controllers (FLCs) enable engineers to develop rule-based systems that mimic human decision-making, making them particularly suitable for nonlinear and uncertain systems (Ghoushchi et al., 2023).

6.5.1.2 Process optimization

Fuzzy logic is instrumental in optimizing engineering processes by considering multiple criteria and constraints:

- **Manufacturing Process Optimization:** In manufacturing, processes often involve multiple variables that need to be controlled to achieve desired outcomes. Fuzzy logic-based control systems adaptively adjust process parameters in response to changing conditions, ensuring robust and efficient operations.
- **Supply Chain Optimization:** Fuzzy logic helps in optimizing supply chain operations by considering factors such as demand variability, lead times, and inventory levels. Fuzzy logic-based models can handle the uncertainty inherent in supply chain dynamics and make real-time decisions to improve efficiency and responsiveness (Rane et al., 2023).

6.5.1.3 Risk assessment and reliability analysis

In engineering, risk assessment and reliability analysis are critical for ensuring the safety and performance of systems:

- **Risk Analysis:** Fuzzy logic is used in risk assessment to evaluate the likelihood and consequences of potential hazards. By incorporating expert knowledge and uncertain data, fuzzy logic models provide a comprehensive assessment of risks, allowing engineers to prioritize mitigation measures effectively.
- **Reliability Engineering:** Fuzzy logic techniques are employed in reliability analysis to assess the performance and lifespan of engineering systems. Fuzzy reliability models can account for uncertainties in failure rates, maintenance schedules, and environmental conditions, enabling engineers to optimize maintenance strategies and improve system reliability (Ibrahim et al., 2023).

In general, fuzzy logic is the backbone of playing MCDM in engineering and design, as it represents an adaptable and intuitive framework for tackling the increasingly complex decision-making challenges within different fields. Because of its capacity to deal with ambiguity, imprecision, and personal preferences, engineers looking for the best answers in various technical application areas will find it very useful. Fuzzy logic offers a broad range of applications in the field of MCDM. This field of study has many applications when decision-making procedures are flexible and intuitive and involve uncertainty and imprecision.

6.5.2 Finance and investment

In the finance sector, where decision-making involves complex risk assessment and uncertainty, fuzzy logic aids in MCDM by

- **Portfolio Management:** Fuzzy logic assists in portfolio selection by considering multiple criteria such as risk, return, and liquidity, accommodating the vagueness and ambiguity inherent in financial data.
- **Credit Risk Assessment:** Fuzzy logic models are employed to assess credit risk by integrating qualitative and quantitative factors, providing a more comprehensive evaluation of borrower creditworthiness.

6.5.3 Environmental management

Environmental management decisions often involve trade-offs between ecological, social, and economic considerations. Fuzzy logic facilitates MCDM in this domain by

- **Environmental Impact Assessment:** Fuzzy logic techniques are utilized to assess the environmental impact of projects or policies by integrating diverse criteria such as biodiversity, habitat loss, and carbon footprint.

- **Waste Management:** Fuzzy logic helps in optimizing waste management strategies by considering criteria such as cost-effectiveness, environmental sustainability, and regulatory compliance (Muhsen et al., 2023).

6.5.4 Healthcare decision-making

In healthcare, where decisions impact patient outcomes and resource allocation, fuzzy logic aids in MCDM by

- **Diagnostic Systems:** Fuzzy logic-based diagnostic systems integrate multiple patient parameters to assist clinicians in decision-making, especially in cases with uncertain or incomplete information.
- **Treatment Planning:** Fuzzy logic models support treatment planning by considering patient preferences, treatment efficacy, and risk factors, enabling personalized and optimized care pathways (Yilmaz, 2023).

6.5.5 Supply chain management

In supply chain management, where decisions involve coordination among multiple stakeholders and optimization of conflicting objectives, fuzzy logic assists in MCDM by:

- **Supplier Selection:** Fuzzy logic techniques are employed to evaluate and select suppliers based on criteria such as quality, cost, reliability, and responsiveness, considering qualitative and quantitative factors.
- **Inventory Management:** Fuzzy logic supports inventory management decisions by dynamically adjusting inventory levels based on demand uncertainty, lead times, and service level requirements, optimizing inventory costs while ensuring customer satisfaction.

These applications demonstrate the versatility and effectiveness of fuzzy logic in addressing complex decision-making challenges across diverse domains, enhancing decision quality, and facilitating better outcomes in MCDM processes (Ala et al., 2023).

6.6 DATA-DRIVEN DECISION-MAKING IN DISASTER MANAGEMENT

Various applications for data-driven decision-making in disaster management are listed as follows:

a. **Early Warning Systems:** Real-time data for early warnings of earthquakes, landslides, floods, volcanoes, tsunamis, wildfires, etc., are acquired through sensors. Data-driven decision-making systems can predict disasters occurring by using AI, Machine Learning (ML), DL, and the Internet of Things (IoT), etc. This makes it possible for communities and government agencies to be prepared in advance and take proactive steps to reduce damage. Figure 6.3 shows categories of early disasters and mitigating technologies for warning. Figure 6.4 shows the predisaster and postdisaster stages of the disaster management cycle.

b. **Resource Allocation:** Distributing essential resources during a disaster as efficiently as possible, such as food, water, medical supplies, and rescue personnel. To determine which places require help the most, various algorithms such as logistic networks are used. This reduces

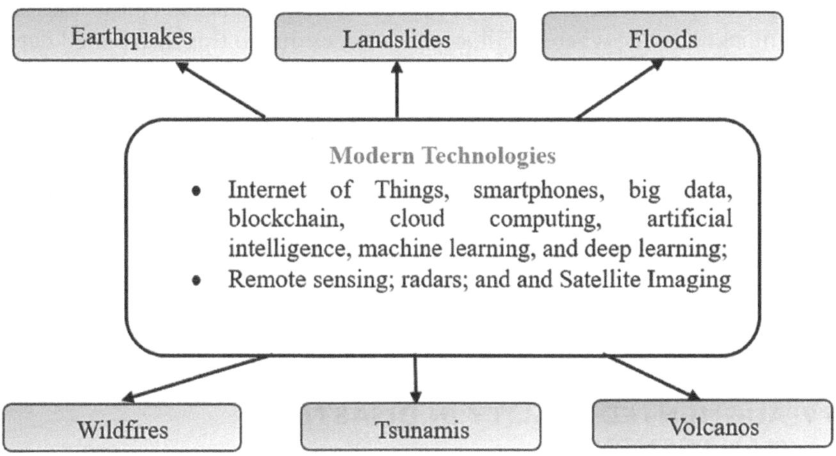

Figure 6.3 Categories of early disaster and mitigating technologies for warning.

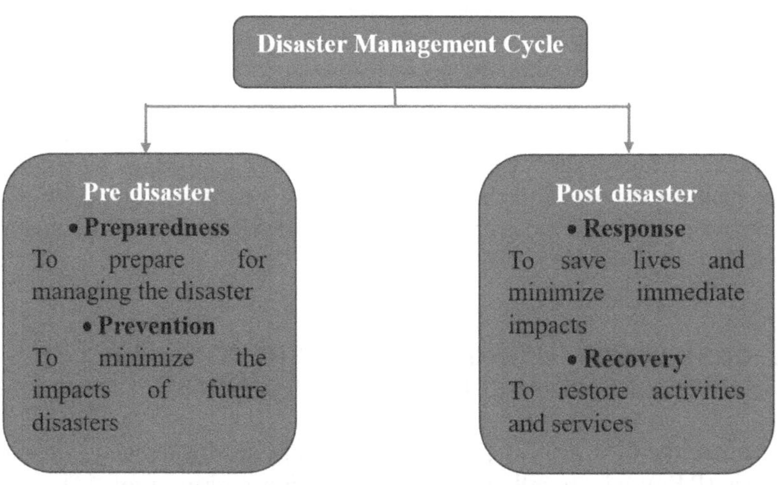

Figure 6.4 Disaster management cycle.

casualties and enhances the effectiveness of relief efforts by ensuring that scarce resources are deployed to the appropriate locations at the appropriate times.

c. **Damage Assessment and Mitigation:** Data-driven decision-making and MCDM systems utilize remote sensing, drone data, and satellite imaging to provide precise, current estimates of the degree of damage. Using this data, reconstruction efforts can be prioritized to guarantee that places that most need assistance get help first.
d. **Collected Data for Disaster Mapping:** Maps of disaster impacts, such as flooded areas, collapsed buildings, and blocked highways, are created in real time by analyzing public data provided by SMS, mobile phones, or social media sites like Twitter and Facebook. Authorities can quickly decide where to allocate resources due to this distributed data.
e. **Disease Outbreak Control and Health Monitoring:** Disease outbreaks are located and monitored through data analysis from hospitals, public reporting, and health records. This data aids in preventing the spread of illness and directing medical resources to the areas most in need of them.
f. **Campaigns for Awareness and Community Involvement:** Public awareness programs are created using risk assessments and data from previous disasters. Local communities manage disasters more effectively when they are informed about possible hazards, evacuation procedures, and emergency response techniques.

6.7 AUGMENTED REALITY IN DISASTER MANAGEMENT

AR technologies play a crucial role in sustainable disaster management by providing real-time information that enhances decision-making across preparedness, response, and recovery phases. AR enables the visualization of hazard zones and evacuation plans during emergencies such as floods and fires, supports earthquake damage assessments, and facilitates immersive climate disaster training simulations. These capabilities help minimize resource wastage while improving overall efficiency and response times. However, the widespread adoption of AR faces challenges such as high implementation costs and the need for reliable infrastructure, particularly in remote or underserved areas. Various uses of AR in disaster management are shown in Figure 6.5.

6.8 CASE STUDY: USING FUZZY LOGIC TO SELECT RENEWABLE ENERGY PROJECTS

To choose the best projects, there is a growing need for renewable energy sources (RES), which calls for efficient decision-making tools. The uncertainty and imprecision that come with assessing RES projects can cause

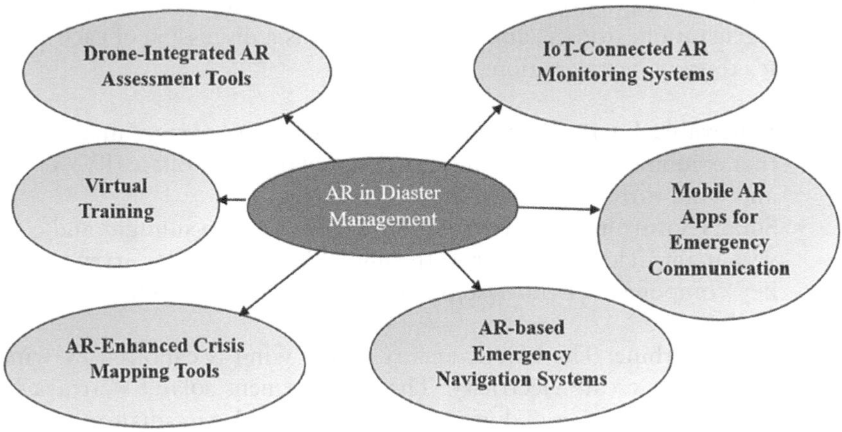

Figure 6.5 Uses of augmented reality in disaster management.

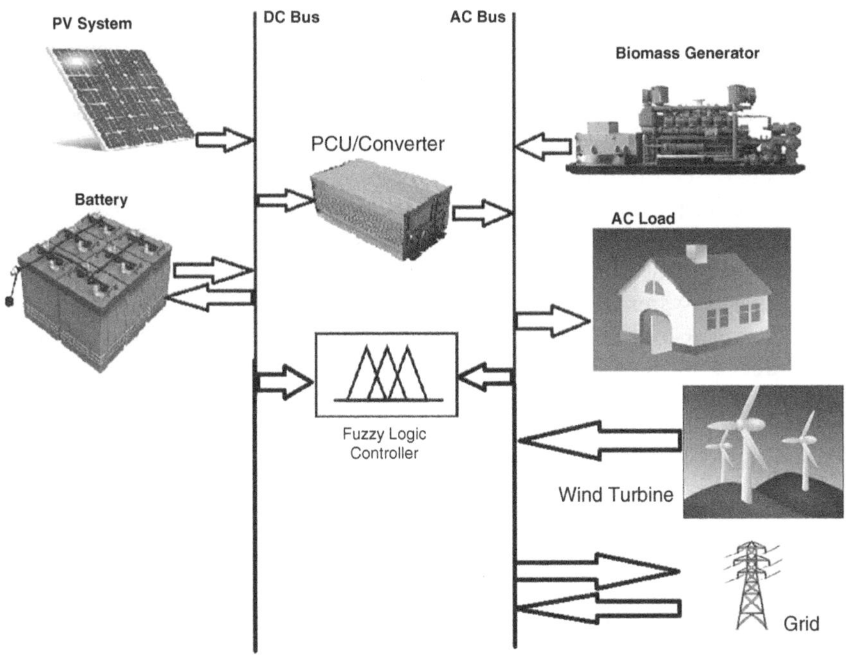

Figure 6.6 Fuzzy logic-based hybrid system.

problems for traditional decision-making techniques. This case study investigates the use of fuzzy logic in the selection of a renewable energy project, taking into account several factors such as social acceptability, environmental effect, and cost (Almasad et al., 2023). A block diagram of a hybrid renewable energy system based on fuzzy logic is shown in Figure 6.6.

It demonstrates how various RES are integrated with an FLC to optimize energy generation, storage, and delivery. Below is a discussion of each component's thorough explanation:

- **Renewable Energy System:** This represents the overarching system that combines various sources, such as solar photovoltaic (PV) arrays and wind turbines, to generate electricity.
- **Solar Photovoltaic Array:** Solar PV arrays capture sunlight and convert it into electrical energy through PV cells. These arrays are a key component of the system, particularly in areas with abundant sunlight.
- **Wind Turbine:** The kinetic energy of the wind is captured by wind turbines to create electricity. They complement solar PV arrays by generating power even during periods of low solar irradiance, such as at night or during cloudy weather.
- **Battery Bank:** As an energy storage device, the battery bank stores extra electricity produced by renewable sources for use at a later time. In times when the production of renewable energy is limited, it helps maintain supply and demand equilibrium by offering a dependable source of electricity.
- **Fuzzy Logic Controller:** At the heart of the system is the FLC, which intelligently manages the operation of RES and battery storage based on input parameters such as energy demand, battery state of charge, and environmental conditions. The FLC utilizes fuzzy logic principles to make decisions dynamically and adaptively, considering the inherent uncertainties and vagueness in renewable energy systems.
- **Power Conditioning Unit/Inverter:** Direct current electricity produced by RES is converted into alternating current electricity that may be used in buildings, commercial spaces, or the electrical grid by the power conditioning unit or inverter. It guarantees that the power produced from renewable resources is suitable for use with the current electrical system.
- **Grid Connection:** Electricity may go back and forth between the renewable energy system and the electrical grid thanks to the grid link. It permits the system to take power from the grid when demand cannot be met by renewable output, and it also permits the system to export excess energy generated to the grid for credit or reimbursement.
- Overall, the fuzzy logic-based hybrid renewable energy system optimizes energy utilization by intelligently controlling the operation of renewable sources and energy storage, thereby maximizing energy efficiency, reliability, and sustainability.

6.9 FUZZY LOGIC IN RENEWABLE ENERGY PROJECT SELECTION

6.9.1 Objectives

- To select the most suitable renewable energy project from a set of alternatives.
- To handle the vagueness and subjectivity in the criteria evaluation.
- To integrate various qualitative and quantitative factors into the decision-making process.

6.9.2 Methodology

6.9.2.1 Criteria definition

- C1, Cost (Initial investment, Operation & maintenance costs)
- C2, Environmental Impact (Carbon emissions, Land use)
- C3, Social Acceptance (Community support, Regulatory approval)
- C4, Technical Feasibility (Energy output, Reliability)

6.9.3 Fuzzy analytic hierarchy process (Fuzzy AHP)

- **Step 1**: Structure the decision hierarchy.
    - **Goal:** Select the best renewable energy project.
    - **Criteria:** Cost, Environmental Impact, Social Acceptance, Technical Feasibility.
    - **Alternatives:** Solar, Wind, Hydro, Biomass.
- **Step 2**: Construct pairwise comparison matrices using fuzzy scales.
    - Use linguistic variables such as "Very High," "High," "Moderate," "Low," and "Very Low" to rate the importance of each criterion relative to others.
- **Step 3**: Derive fuzzy weights for each criterion.
    - Convert linguistic comparisons into fuzzy numbers.
    - Compute the fuzzy weight for each criterion using the geometric mean method.
- **Step 4**: Evaluate alternatives against each criterion using fuzzy ratings.
    - Rate each alternative for each criterion using fuzzy scales.
    - Aggregate the ratings to obtain a fuzzy performance score for each alternative.

6.9.4 Steps in fuzzy logic in MCDM

Step 1: Create the Fuzzy Pairwise Comparison Matrix using the formula (Eq. 6.1) provided.

Regarding the cost, environmental impact, social acceptability, and technical feasibility (C1, C2, C3, and C4) criteria:

$$B = \begin{bmatrix} (1,1,1) & b_{12} & b_{13} & b_{14} \\ \frac{1}{b_{12}} & (1,1,1) & b_{23} & b_{24} \\ \frac{1}{b_{13}} & \frac{1}{b_{23}} & (1,1,1) & b_{34} \\ \frac{1}{b_{14}} & \frac{1}{b_{24}} & \frac{1}{b_{34}} & (1,1,1) \end{bmatrix} \tag{6.1}$$

where

b_{ij} are fuzzy numbers representing the relative importance of criteria i and j.

Normalization adjusts the values to a comparable scale.

Step 2: Compute the fuzzy weights using the Geometric Mean Method by using Eq. (6.2).

For each criterion C_i:

$$w_{i=}\left(\prod_{j=1}^{n} a_{ij}\right)^{\frac{1}{n}} \tag{6.2}$$

Normalize the fuzzy weights using Eq. (6.3):

$$w_i = \frac{w_i}{\sum_{i=1}^{n} w_i} \tag{6.3}$$

Step 3: Using Eq. (6.4), create the Fuzzy Decision Matrix for Alternatives.

For alternatives A1, A2, A3, and A4 (solar, wind, hydro, biomass):

$$FD = \begin{bmatrix} fd_{11} & fd_{12} & fd_{13} & fd_{14} \\ fd_{21} & fd_{22} & fd_{23} & fd_{24} \\ fd_{31} & fd_{32} & fd_{33} & fd_{34} \\ fd_{41} & fd_{42} & fd_{43} & fd_{44} \end{bmatrix} \tag{6.4}$$

where

fd_{ij} are fuzzy numbers representing the performance of alternative i w.r.t. criterion j.

Step 4: Determine fuzzy positive ideal solution (FPIS) and fuzzy negative ideal solution (FNIS) using Eqs. (6.5 and 6.6).

FPIS (A^+) and FNIS (A^-)

$$A^+ = \left(\max d_{ij}\right) \tag{6.5}$$

$$A^- = \left(\min d_{ij}\right) \tag{6.6}$$

Step 5: Calculate the distance from FPIS and FNIS by using Eqs. (6.7 and 6.8).

For each alternative A_i:

$$D_I^+ = \sqrt{\sum_{J=1}^{m} \left(d_{ij} - A_J^+\right)^2} \tag{6.7}$$

$$D_I^- = \sqrt{\sum_{J=1}^{m} \left(d_{ij} - A_J^-\right)^2} \tag{6.8}$$

Step 6: Compute the closeness coefficient by using Eq. (6.9).

$$CC_i = \frac{D_i^-}{D_i^+ + D_i^-} \tag{6.9}$$

Step 7: Sort and order the options. Selecting the option with the highest closeness coefficient (CCi) is the optimal decision.

Step 8: Fuzzy weights.

Using the fuzzy weights and fuzzy decision matrix, we can derive the final scores for each alternative. For illustration purposes, consider the following normalized fuzzy weights and decision matrix as shown in Eq. (6.10):

$w_{cost=}(0.2, 0.3, 0.4)$

$w_{env=}\ (0.1, 0.2, 0.3)$

$w_{social} = (0.3, 0.4, 0.5)$

$w_{technical} = (0.2, 0.3, 0.4)$

$$D = \begin{bmatrix} (0.6,0.7,0.8) & (0.5,0.6,0.7) & (0.3,0.4,0.5) & (0.5,0.6,0.7) \\ (0.7,0.8,0.9) & (0.5,0.6,0.7) & (0.4,0.5,0.6) & (0.6,0.7,0.8) \\ (0.8,0.9,1.0) & (0.6,0.7,0.8) & (0.5,0.6,0.7) & (0.7,0.8,0.9) \\ (0.9,1.0,1.1) & (0.7,0.8,0.9) & (0.6,0.7,0.8) & (0.8,0.9,1.0) \end{bmatrix} \tag{6.10}$$

Calculate closeness coefficient (CC):

$$CC_{Solar=}0.85,\ CC_{Wind=}0.75,\ CC_{Hydro=}0.65,3\ CC_{Biomass=}0.80.$$

The use of fuzzy logic in the selection of a renewable energy project allows the inclusion of many criteria, often conflicting with each other; it also manages the uncertainties inherent in the evaluation procedure. In the case study, it is proven that solar is the best among the alternatives considered since it has the highest CC. This methodology would also permit approaches to decision-making that are systematic and flexible, as all relevant factors would be considered and then accurately weighed, which would mean more robust and informed decisions for the renewable energy sector.

6.10 CHALLENGES AND FUTURE DIRECTIONS IN FUZZY LOGIC-BASED MCDM

6.10.1 Challenges

Fuzzy logic-based MCDM faces several challenges that impede its widespread adoption and effectiveness across various domains. One of the foremost challenges is the computational complexity associated with fuzzy logic-based methods. As decision problems become more intricate with a larger number of criteria and alternatives, the computational burden escalates, demanding significant resources and time for processing. This complexity inhibits the scalability of fuzzy MCDM, limiting its applicability to smaller-scale problems. Additionally, fuzzy MCDM relies heavily on subjective assessments and expert opinions to define linguistic variables and membership functions. This subjectivity introduces uncertainties and biases into the decision-making process, potentially leading to unreliable outcomes. Moreover, traditional fuzzy MCDM methods often assume static preferences and criteria, failing to account for dynamic environments where preferences and criteria evolve. This limitation hampers the adaptability of fuzzy MCDM models to changing circumstances, resulting in suboptimal decisions (Więckowski et al., 2023).

Another set of difficulties arises from integration with developing technologies, such as interoperability, security, and data privacy concerns. Strong efforts are needed to overcome these issues and guarantee platform compatibility to integrate fuzzy MCDM with technologies like AI, ML, and the IoT in a smooth manner. Furthermore, there are issues with fuzzy MCDM models' interpretability and transparency as their complexity might make it difficult for decision-makers to comprehend the reasoning behind the outcomes that are produced.

The inability to be easily understood might make people skeptical and reluctant to use fuzzy MCDM techniques. Future research efforts must

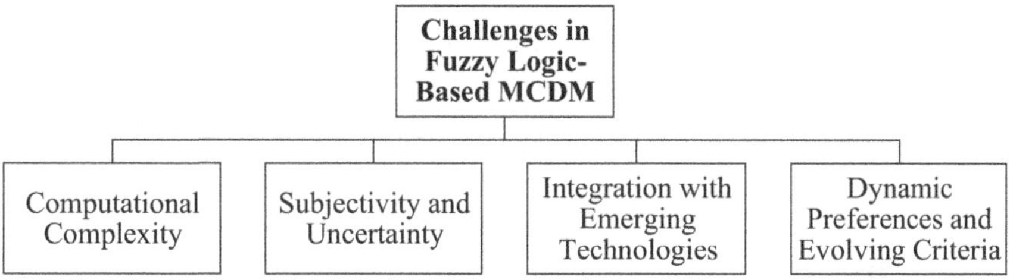

Figure 6.7 Challenges in fuzzy logic-based MCDM.

concentrate on creating more effective algorithms to decrease computational complexity, improving techniques for combining expert opinions to reduce subjectivity, and incorporating dynamic elements into fuzzy MCDM models to account for changing criteria and preferences to address these issues.

Moreover, efforts to ensure data privacy, security, and interoperability in the integration with emerging technologies are essential. Improving model interpretability through visualization techniques and stakeholder involvement can enhance the trust and acceptance of fuzzy MCDM methods. By addressing these challenges (Figure 6.7), fuzzy logic-based MCDM can realize its full potential as a powerful tool for decision-making in complex and uncertain environments (Castillo-Calzadilla et al., 2023; Bairagi, 2023).

6.10.2 Future directions

Future directions in fuzzy logic-based MCDM hold significant promise for advancing decision-making processes in complex and uncertain environments. One crucial area of exploration involves enhancing computational efficiency through the development of optimized algorithms and the utilization of parallel and distributed computing techniques. These efforts aim to reduce computational complexity and improve the scalability of fuzzy MCDM methods, enabling their application to larger-scale decision problems. Additionally, addressing subjectivity and uncertainty remains a key focus, with research aiming to refine consensus-building techniques and develop advanced sensitivity analysis methods to ensure robust and reliable decision outcomes (Gribiss et al., 2023). Furthermore, the evolution toward dynamic and adaptive decision models involves exploring dynamic fuzzy sets and real-time decision support systems that can adapt to changing preferences and environmental conditions. Fuzzy MCDM's capabilities might be enhanced by integration with cutting-edge technologies like blockchain, AI, and ML, opening the door to the creation of hybrid decision-making frameworks that are more effective and efficient. To translate these developments into workable answers for real-world decision-making problems, interdisciplinary cooperation and implementation across diverse industry sectors are crucial. Through the pursuit of these forthcoming avenues, scholars might additionally liberate the capacity of fuzzy logic-based MCDM to tackle intricate decision-making challenges and stimulate creativity throughout various fields (Da Silva Romero et al., 2023; Ghai et al., 2022).

6.11 CONCLUSION

Fuzzy logic-based MCDM techniques provide effective tools for handling complex decision-making in uncertain and evolving environments. By incorporating fuzzy logic principles, these techniques offer a flexible framework for evaluating alternatives across multiple criteria while addressing ambiguity and uncertainty. As highlighted in this study, fuzzy MCDM has significant potential in areas such as supply chain optimization, environmental management, engineering design, finance, and healthcare. However, several challenges must be addressed to maximize its effectiveness. Computational complexity hampers scalability, necessitating the development of efficient algorithms and advanced computing techniques. Additionally, managing subjectivity and uncertainty remains a priority, with ongoing efforts focused on refining consensus-building approaches and incorporating sensitivity analysis to improve decision reliability. The integration of fuzzy MCDM with emerging technologies such as AI, ML, and blockchain offers opportunities to create hybrid decision-making frameworks with enhanced efficiency. Future research in fuzzy MCDM will focus on improving computational efficiency, addressing subjectivity, developing dynamic decision models, and leveraging advanced technologies. Interdisciplinary collaboration and real-world applications across various industries will be essential in translating theoretical advancements into practical solutions. Ultimately, fuzzy logic-based MCDM serves as a valuable approach for navigating complex decisions in dynamic settings. By embracing continuous research and technological innovations, practitioners can leverage these methods to make informed decisions, foster innovation, and achieve sustainable outcomes. As fuzzy MCDM continues to evolve, its role in shaping the future of decision-making will grow, contributing to more effective and resilient decision processes.

REFERENCES

Abdul, D., Wenqi, J., Tanveer, A., & Sameeroddin, M. (2024). Comprehensive analysis of renewable energy technologies adoption in remote areas using the integrated delphi-fuzzy AHP-VIKOR approach. *Arabian Journal for Science and Engineering*, *49*(5), 7585–7610.

Arora, K., Joshi, G. P., Ragab, M., Rawa, M., Milyani, A. H., Mansour, R. F., & Yang, E. (2022). Bilateral contract for load frequency and renewable energy sources using advanced controller. *Computers, Materials & Continua*, *73*(2), 3165–3180. https://doi.org/10.32604/cmc.2022.026966

Aghaloo, K., Ali, T., Chiu, Y. R., & Sharifi, A. (2023). Optimal site selection for the solar-wind hybrid renewable energy systems in Bangladesh using an integrated GIS-based BWM-fuzzy logic method. *Energy Conversion and Management*, *283*, 116899.

Afolabi, T., & Farzaneh, H. (2023). Optimal design and operation of an off-grid hybrid renewable energy system in Nigeria's rural residential area, using fuzzy logic and optimization techniques. *Sustainability*, *15*(4), 3862.

Ala, A., Mahmoudi, A., Mirjalili, S., Simic, V., & Pamucar, D. (2023). Evaluating the performance of various algorithms for wind energy optimization: A hybrid decision-making model. *Expert Systems with Applications*, *221*, 119731.

Alanazi, A., & Alanazi, M. (2023). Multicriteria decision-making for evaluating solar energy source of Saudi Arabia. *Sustainability*, *15*(13), 10228.

Ali, G., Musbah, H.N., Aly, H.H., & Little, T. (2023). Hybrid renewable energy resources selection based on multi criteria decision methods for optimal performance. *IEEE Access*, *11*, 26773–26784.

Almasad, A., Pavlak, G., Alquthami, T., & Kumara, S. (2023). Site suitability analysis for implementing solar PV power plants using GIS and fuzzy MCDM based approach. *Solar Energy*, *249*, 642–650.

Bairagi, B. (2023). A fuzzy interval based multi-criteria homogeneous group decision making technique: An application to airports ranking problem. *Decision Making: Applications in Management and Engineering*, *6*(2), 1–15.

Beheshtikhoo, A., Pourgholi, M., & Khazaee, I. (2023). Design of type-2 fuzzy logic controller in a smart home energy management system with a combination of renewable energy and an electric vehicle. *Journal of Building Engineering*, *68*, 106097.

Castillo-Calzadilla, T., Garay-Martinez, R., & Andonegui, C. M. (2023). Holistic fuzzy logic methodology to assess positive energy district (PathPED). *Sustainable Cities and Society*, *89*, 104375.

Channi, H.K., Sandhu, R., Faiz, M., & Islam, S.M. (2023, August). Multi-criteria decision-making approach for laptop selection: A case study. In: *2023 3rd Asian Conference on Innovation in Technology (ASIANCON),* Ravet IN, India (pp. 1–5). IEEE.

Da Silva Romero, C.W., Miyazaki, M.R., Berni, M.D., Figueiredo, G.K.D.A., & Lamparelli, R.A.C. (2023). A spatial approach for integrating GIS and fuzzy logic in multicriteria problem solving to support the definition of ideal areas for biorefinery deployment. *Journal of Cleaner Production*, *390*, 135886.

Dutta, R., Das, S., & De, S. (2023). Multi criteria decision making with machine-learning based load forecasting methods for techno-economic and environmentally sustainable distributed hybrid energy solution. *Energy Conversion and Management*, *291*, 117316.

Faiz, M., & Daniel, A.K. (2024). A multi-criteria cloud selection model based on fuzzy logic technique for QoS. *International Journal of System Assurance Engineering and Management*, *15*(2), 687–704.

Garcia-Orozco, S., Vargas-Gutierrez, G., Ordonez-Sanchez, S., & Silva, R. (2023). Using multi-criteria decision making in quality function deployment for offshore renewable energies. *Energies*, *16*(18), 6533.

Ghai, D., Tripathi, S.L., Saxena, S., Chanda, M., & Alazab, M. (Eds.). (2022). *Machine Learning Algorithms for Signal and Image Processing*. John Wiley & Sons.

Ghoushchi, S.J., Garg, H., Bonab, S.R., & Rahimi, A. (2023). An integrated SWARA-CODAS decision-making algorithm with spherical fuzzy information for clean energy barriers evaluation. *Expert Systems with Applications*, *223*, 119884.

Göçer, F. (2024). A novel extension of Fermatean fuzzy sets into group decision making: A study for prioritization of renewable energy technologies. *Arabian Journal for Science and Engineering*, *49*(3), 4209–4228.

Gribiss, H., Aghelinejad, M.M., & Yalaoui, F. (2023). Configuration selection for renewable energy community using MCDM methods. *Energies*, *16*(6), 2632.

Hussain, A., Ullah, K., Pamucar, D., Haleemzai, I., & Tatić, D. (2023). Assessment of solar panel using multiattribute decision-making approach based on intuitionistic fuzzy Aczel Alsina Heronian mean operator. *International Journal of Intelligent Systems*, *2023*(1), 6268613.

Ibrahim, O., Bakare, M.S., Amosa, T.I., Otuoze, A.O., Owonikoko, W.O., Ali, E.M., … & Ogunbiyi, O. (2023). Development of fuzzy logic-based demand-side energy management system for hybrid energy sources. *Energy Conversion and Management: X*, *18*, 100354.

Kumar, P., & Channi, H.K. (2024, February). Optimal electric vehicle charging station placement: A multi-criteria decision-making approach for site selection. In: *2024 International Conference on Integrated Circuits and Communication Systems (ICICACS)*, Raichur, India (pp. 1–6). IEEE.

Kumar, R., Channi, H.K., & Singh, H. (2020). Selection of mobile phone with multi criteria decision making approach: A case study. In: *Future of Business through Innovations,* Ludhiana, Punjab, India.

Khoshsepehr, Z., Alinejad, S., & Alimohammadlou, M. (2023). Exploring industrial waste management challenges and smart solutions: An integrated hesitant fuzzy multi-criteria decision-making approach. *Journal of Cleaner Production*, *420*, 138327.

Lotfi, R., Gharehbaghi, A., Mehrjardi, M.S., Kheiri, K., & Ali, S. S. (2023). A robust, resilience multi-criteria decision-making with risk approach: A case study for renewable energy location. *Environmental Science and Pollution Research*, *30*(15), 43267–43278.

Martínez, L., Dinçer, H., & Yüksel, S. (2023). A hybrid decision making approach for new service development process of renewable energy investment. *Applied Soft Computing*, *133*, 109897.

Manickam, P., Girija, M., Dutta, A. K., Ramesh Babu, P., Arora, K., Jeong, M. K., & Acharya, S. (2024). Empowering cybersecurity using enhanced rat swarm optimization with deep stack-based ensemble learning approach. *IEEE Access*. https://doi.org/10.1109/ACCESS.2024.3395328

Muhsen, Y.R., Husin, N.A., Zolkepli, M.B., Manshor, N., & Al-Hchaimi, A.A. J. (2023). Evaluation of the routing algorithms for NoC-based MPSoC: A fuzzy multi-criteria decision-making approach. *IEEE Access, 11*, 102806–102827.

Rane, N., Achari, A., & Choudhary, S. (2023). Multi-criteria decision-making (MCDM) as a powerful tool for sustainable development: Effective applications of AHP, FAHP, TOPSIS, ELECTRE, and VIKOR in sustainability. *International Research Journal of Modernization in Engineering Technology and Science*, *5*(4), 2654–2670.

Razeghi, M., Hajinezhad, A., Naseri, A., Noorollahi, Y., & Moosavian, S.F. (2023). Multi-criteria decision-making for selecting a solar farm location to supply energy to reverse osmosis devices and produce freshwater using GIS in Iran. *Solar Energy*, *253*, 501–514.

Sahoo, S.K., & Goswami, S.S. (2023). A comprehensive review of multiple criteria decision-making (MCDM) Methods: Advancements, applications, and future directions. *Decision Making Advances*, *1*(1), 25–48.

Sahu, K., Srivastava, R.K., Kumar, S., Saxena, M., Gupta, B.K., & Verma, R.P. (2023). Integrated hesitant fuzzy-based decision-making framework for evaluating sustainable and renewable energy. *International Journal of Data Science and Analytics*, *16*(3), 371–390.

Štilić, A., & Puška, A. (2023). Integrating multi-criteria decision-making methods with sustainable engineering: A comprehensive review of current practices. *Eng*, *4*(2), 1536–1549.

Taherdoost, H., & Madanchian, M. (2023). Multi-criteria decision making (MCDM) methods and concepts. *Encyclopedia*, *3*(1), 77–87.

Tafula, J.E., Justo, C.D., Moura, P., Mendes, J., & Soares, A. (2023). Multicriteria decision-making approach for optimum site selection for off-grid solar photovoltaic microgrids in Mozambique. *Energies*, *16*(6), 2894.

Tian, G., Lu, W., Zhang, X., Zhan, M., Dulebenets, M.A., Aleksandrov, A., ... & Ivanov, M. (2023). A survey of multi-criteria decision-making techniques for green logistics and low-carbon transportation systems. *Environmental Science and Pollution Research*, *30*(20), 57279–57301.

Więckowski, J., Kizielewicz, B., Shekhovtsov, A., & Sałabun, W. (2023). How do the criteria affect sustainable supplier evaluation? A case study using multi-criteria decision analysis methods in a fuzzy environment. *Journal of Engineering Management and Systems Engineering*, *2*(1), 37–52.

Yilmaz, I. (2023). A hybrid DEA–fuzzy COPRAS approach to the evaluation of renewable energy: A case of wind farms in Turkey. *Sustainability*, *15*(14), 11267.

Zahid, K., & Akram, M. (2023). Multi-criteria group decision-making for energy production from municipal solid waste in Iran based on spherical fuzzy sets. *Granular Computing*, *8*(6), 1299–1323.

Zaliluddin, D. (2023). Bibliometric analysis of "accuracy of multi criteria decision making (MCDM) of assistance recipients with fuzzy logic algorithm". *West Science Interdisciplinary Studies*, *1*(7), 353–363.

Chapter 7

Conceptualizing the access and usage of AI as a fundamental right for netizens

Muhammad Younus, Halimah Abdul Manaf, Achmad Nurmandi, Andi Adawiah, Dyah Mutiarin, Eliza Meiyani, Titin Purwaningsih, Ulung Pribadi, Imron Sohsan, Andi Luhur Prianto, and Sandi Lubis

7.1 INTRODUCTION

Artificial intelligence (AI) is now an intrinsic part of modern society, fully revolutionizing how people interact, access information, and conduct everyday activities. Due to its omnipresence, the classification of AI as a human right has been the subject of hot debate. This article critically analyzes the notion of conceptualizing AI usage and access as a fundamental human right for netizens and explores the implications, challenges, and prospects of this framing. The beneficial use of AI access is increasingly understood as a necessary requirement for human beings to exercise and enjoy their inalienable human rights. Through its multivaried reports and resolutions, the United Nations has emphatically reiterated the significance of access to AI in ensuring the availability of rights like freedom of expression, the right to information, and meaningful participation in cultural, social, and political life. Additionally, the AI has a significant role to play in the enjoyment of economic and social rights like but not limited to the right to education, health, and employment. Thus, the notion of conceptualizing AI access and use as a basic human right is not only consistent with current international human rights frameworks but also necessary to guarantee the full realization of people's rights in the digital world.

Nonetheless, despite the increasing consciousness regarding the access to AI being an essential need, there remain significant disparities with regard to its availability and affordability across different regions and societies.

The digital divide, characterized by asymmetrical access to and utilization of digital technologies, exacerbates prevalent socioeconomic inequalities and bars marginalized groups from fully benefiting from their human rights. Closing this digital gap will have to include collaborative efforts by the governments, private sector entities, civil society, and the global community as a whole in a bid to extend universal access to cheap and reliable

 DOI: 10.1201/9781003641582-7

AI infrastructure and services. In addition, framing AI access and utilization as an elementary human right introduces complex legal, ethical, and policy challenges. Although most proponents make the case for AI access being a human right due to its pivotal role in facilitating the enjoyment of other rights, there are also detractors who warn against such a proclamation, citing possible implications on AI regulation, freedom of expression, and the consolidation of privacy rights. Besides, the transnational nature of AI presents serious challenges in creating a harmonious legal regime that can safeguard and promote the rights of individuals in cyberspace effectively. Meeting these challenges effectively requires the utilization of a multidisciplinary approach that interlinks with the competence of ensuring global access to AI, safeguarding the rights of individuals, and enhancing digital literacy and online safety.

Governments also have an important role to play to ensure an environment for universal AI access by making investments in infrastructure development, embracing visionary policies that offer affordability and competitiveness in the telecommunication industry, and protecting the rights of internet users by ensuring solid legal frameworks and accountability mechanisms. It is also important to realize the worth of efforts toward developing digital literacy and enabling people to go online securely and responsibly. Efforts toward making the complete potential of AI access a universal human right are essential. At the center of this is the active participation of civil society organizations, academia, and the private sector. They accomplish this through carrying out advocacy, through capacity-building efforts, and creating tools and resources that enhance digital inclusion and empower citizens to exercise their rights online. In summary, it cannot be denied that AI has turned into an indispensable tool in achieving core human rights in the digital era.

Enfranchising AI access and use as a human right is critical to closing gaps, ensuring inclusivity, and allowing people to enjoy their rights online in their entirety. However, realizing this vision requires positive and collective action by governments, civil society, and the private sector. Such action must be aimed at closing the digital divide, protecting users' rights, and fostering digital literacy and empowerment. This study seeks to add to the discussion by critically analyzing the implications, risks, and potentialities of framing AI access as a universal human right.

7.2 LITERATURE REVIEW

With its pervasive existence in virtually every area of contemporary life, AI has produced a radical paradigm shift in communication, learning, commerce, and information availability. The availability of AI has raised the issue of an interesting and crucial debate regarding whether access to and use of the AI can or cannot be referred to as a human right (Chemmencheri, 2016).

In an attempt to shed some light on the issue, the current paper reviews the literature available, weighing the arguments for and against this statement in a balanced way, and assessing the different schools of thought which have been embraced in examining the different roles played by AI in promoting human rights (Manickam et al., 2024).

Conversely, the advocates of the view that sees the AI as a human right hold the view that it is an amplifier of the exercise of already established fundamental human rights enshrined in international declarations. According to them, the AI offers instant and cheap information on people's fingertips that is necessary for people to be able to fully and intelligently engage in a democratic society (Amodu et al., 2019). Also, it is claimed that the AI enables people by granting them the freedom to speak and participate in public debate (Chen, 2015). Also, it is claimed that the AI is an educational site, by which people have access to information and enable individual growth (Büchi, 2016). Moreover, the AI is also used as a driver of economic engagement through channels like e-commerce and online entrepreneurial opportunities, which ultimately help in alleviating poverty and enhancing living standards (Llorente-Barroso et al., 2015). Denying access to AI in this regard would be tantamount to denying people the means by which they can actualize these basic rights.

Conversely, critics of the "human right to access AI" maintain that it is an overstatement. They invoke the intrinsic limitation of this precept and underscore that access to the AI is not necessarily a guarantee of proper utilization (Arora et al., 2022a). They draw attention to possible constraints that interfere with appropriate utilization, including poor digital literacy skills, culture, and infrastructure challenges (Wang & Liu, 2018). Besides, critics point out the potential risk and harm that could be generated by way of the internet community, including disinformation, hate speech, and cybercrime, which raise issues concerning the need to exercise responsible use and whether or not to establish restrictions on content (Arora et al., 2022b). In addition, issues regarding net neutrality and data protection also undermine the access control and quality of AI, indicating that even access may not be enough (Mascheroni & Ólafsson, 2016).

Outside these two poles, numerous researchers have presented balanced conceptual models in an effort to promote our understanding of the interaction between human rights and AI. One of these discusses the "right to connectivity," which acknowledges the importance of access but brings a similar emphasis to the quality and intent of that access (Choudrie et al., 2010). This type of approach facilitates cheap, stable, and open AI links, allowing people to interact effectively with the virtual community. Another model, or "digital inclusion," identifies economic and social disparity as a barrier to use of the AI and also requires giving people the appropriate skills in order to take full advantage of the AI (Arora et al., 2021). This plan calls for taking on policies to bridge the digital divide and facilitate equal access to the benefits brought by the AI.

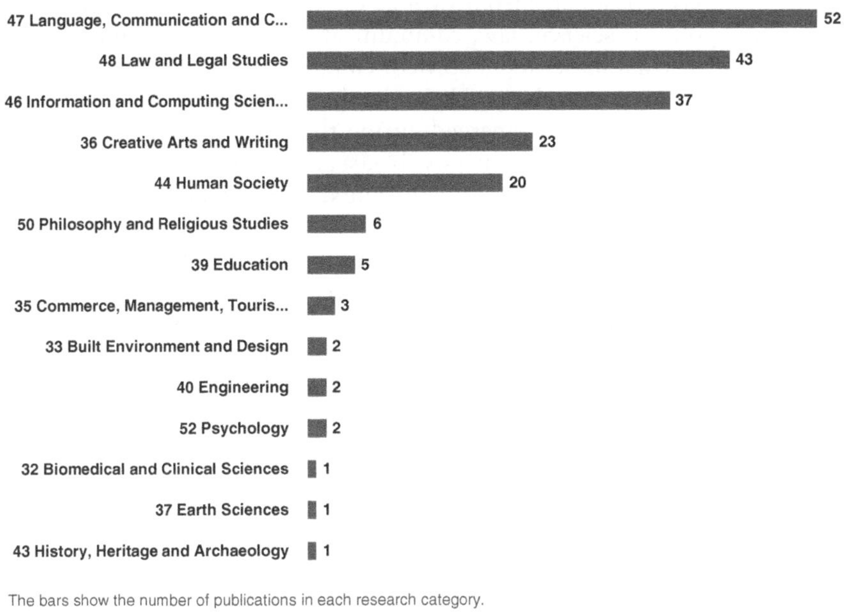

Figure 7.1 Publication by subject areas.

Also, human rights standards have been suggested specifically for the online space. For example, the Santa Clara Principles suggest encouraging freedom of expression, access to information, and privacy on the internet while recognizing the necessity of regulations to protect these rights in the context of the current online issues (Singh et al., 2024) (Figure 7.1).

7.3 RESEARCH METHOD

This research utilizes a qualitative method that is typified by an extensive literature study. The research aims to research and conceptualize the use and access of the AI as an inherent human right of netizens. The justification for using the qualitative method lies in the fact that it naturally has the potential to explore the complexities and sensitivities of the topic in-depth. This method facilitates a rich comprehension of AI access and uses complicated perspectives, experiences, and settings. Combining the existing literature, this research attempts to present a fair overview of the new debate about AI access as a human right. This will uncover principal themes, arguments, and implications of this debate, finally for purposes of policy and practice.

Literature review is an organized process of searching, selecting, and critically reviewing relevant scholarly articles, books, reports, and policy papers. Material originates from various disciplines though not limited to sociology, political science, law, communication studies, and information technology. The databases like PubMed, Google Scholar, JSTOR, and Web of Science are used effectively. These databases are good sources to find relevant literature that has emerged during the past two decades. To provide as comprehensive a view as possible, search terms are employed, e.g., "AI access," "digital divide," "human rights," "information society," and corresponding keywords.

In the instance of literature selection criteria, emphasis is placed on studies dealing with conceptualizing, importance, problems, and implications of AI access as a human right. These studies are taken from multiple theoretical, empirical, and methodological underpinnings. Exclusion criteria, in contrast, are irrelevant or old sources, non-peer-reviewed publications, and duplicates. The first search results in a large amount of literature, which is screened systematically with regard to its applicability to the research questions and thematic focus. The screened literature is synthesized critically and discussed to determine overall themes, theoretical concepts, and empirical findings related to conceptualizing AI access as a human right. Coding and sorting data are processes done to determine patterns, contradictions, and gaps in the literature. In addition to this analysis, synthesis involves critical readings of epistemological and ideological origins that frame debates regarding AI accessibility and human rights. These involve a reading of power relations, socioeconomic disparities, cultures, and perceptions of technological determinism (Figure 7.2).

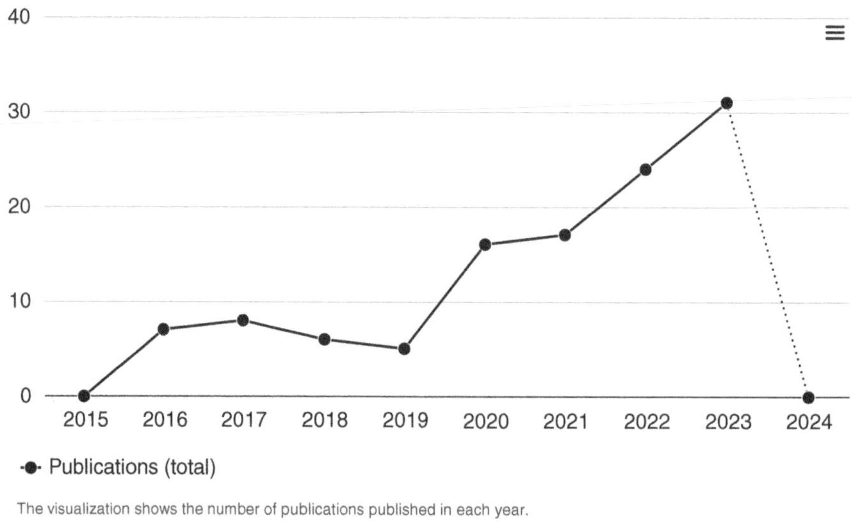

Figure 7.2 Publication by years.

7.4 RESULTS AND DISCUSSION

In this section, we shall put forth the outcomes of our scholarly inquiry concerning the conceptualization of the access and utilization of AI as an essential entitlement bestowed upon individuals, followed by a comprehensive examination of the ramifications of these discoveries.

7.4.1 Results

7.4.1.1 International Consensus on the Right to Access AI

Our comprehensive analysis has revealed an emerging international consensus among policymakers, international institutions, and civil society actors that categorically endorse access to AI as an inalienable and inherent human right. This new consensus is strongly reaffirmed and entrenched by a range of international statements, resolutions, and agreements that emphatically reiterate the highest priority of ensuring universal access to AI, thereby making it a sine qua non element of modern human rights rhetoric.

7.4.1.2 Legal Recognition

It has been legally acknowledged by the majority of the world's jurisdictions that the right of access to AI has been legally established and enshrined through constitutional provisions, enactments, or milestone judicial pronouncements. These carefully drafted legal principles all reunite the central role played by the AI in making the exercise of other basic rights, including but not limited to freedom of expression, education, and active engagement in public life, effective. By embracing the law of AI access, these states have reaffirmed the importance of ensuring that individuals are facilitated to engage with the digital world fully and effectively.

7.4.1.3 Barriers to Access

While praiseworthy efforts in acknowledging the right of access to AI, there are daunting barriers still deterring the attainment of global connectivity. These complex challenges include a number of issues, including but not limited to infrastructure constraints, including but not limited to the glaring lack of broadband reach in rural and remote communities, as well as socioeconomic hurdles, namely the entrenched issues of affordability and the pervasive digital literacy gaps. Solving these challenges in full and in an equitable manner is paramount to making universal AI accessible and the resultant empowerment of citizens and communities.

7.4.1.4 Digital Inclusion Initiatives

Our extensive research has meticulously recognized and examined many digital inclusion initiatives that have responsibly worked toward bridging the access gap and forging more inclusive and equitable participation in the online environment. Such cutting-edge initiatives involve a broad spectrum of approaches, from government initiatives to community initiatives and dynamic cooperation among the private sector and like-minded stakeholders. These efforts are primarily directed toward scaling and building AI infrastructure, enabling low-cost hardware provision, and offering full and balanced digital skills training to historically excluded and underserved individuals and groups.

7.4.1.5 Influence on Human Development

The access to the AI has a most profound and revolutionary effect on all aspects of human development, ranging from education, health, economic opportunity, to civic participation. Our stringent examination and incisive analysis of empirical evidence obviously establish that broadening access to AI can make a meaningful contribution to poverty alleviation, fostering better health outcomes and more civic participation, ultimately leading to a more equitable and inclusive society (Figure 7.3).

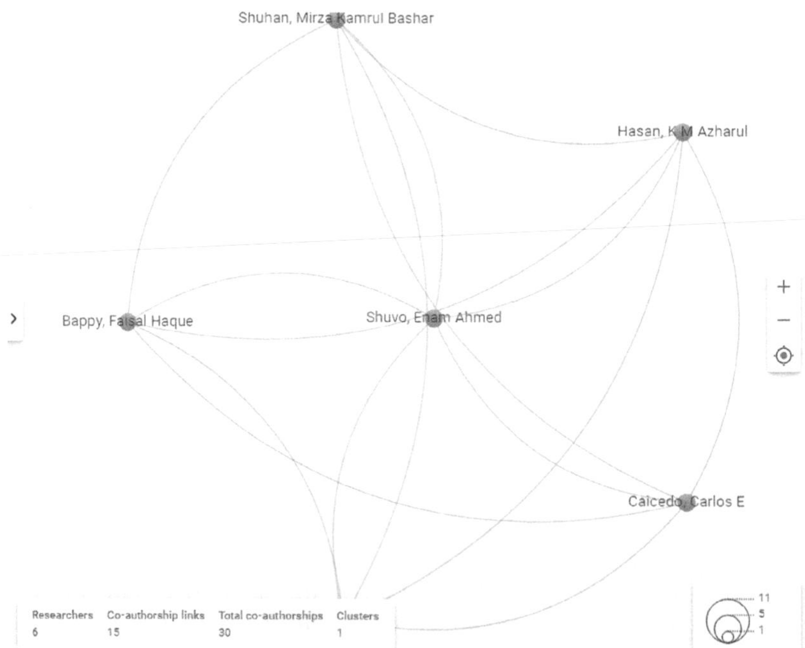

Figure 7.3 Co-citation analysis of authors.

7.4.2 Discussion

7.4.2.1 Intersectionality of Rights

The self-conscious recognition of AI access as a fundamental human right qualitatively highlights the complex and intertwined nature of civil, political, economic, social, and cultural rights. The AI, being a communication and information tool with immense potential, is central to helping people enjoy their rights to freedom of expression, information, assembly, and association without any setback and efficiently while at the same time contributing to the facilitation of the realization of other numerous rights, including the right to work, education, and cultural activities. Recognizing and developing the intersectionality of these rights is key to the overall and comprehensive development of human rights in the digital age.

7.4.2.2 Digital Divide

Amidst progress in spreading AI accessibility across the world, there are widening gaps across fault lines such as geographic, socioeconomic, and demographic characteristics. Overcoming the long-standing digital divide effectively entails starting a holistic and integrative strategy to overcome infrastructural deficiencies and confront costs and literacy in a simple manner. Besides this, making the participation and empowerment of marginal and vulnerable groups, who have additional barriers to AI access and digital inclusion, is essential.

7.4.2.3 Regulatory Frameworks

Strong regulatory frameworks are essential to maintaining the right to access to AI and building a digital world characterized by freedom, openness, and inclusiveness. Such regulatory frameworks will need to thread a thin line between the inevitable need to trigger competition and innovation and the equally vital function of protecting user rights, privacy, and digital security. Through the adoption of a robust and well-structured regulatory environment, the governments can harness the potential of AI efficiently while, at the same time, offering the people protection from potential harm and exploitation.

7.4.2.4 Public–Private Partnerships

To achieve successful progress in digital inclusion goals, there is an urgent need to develop collaborative and cooperative partnerships among governments, civil society, academia, and the private sector. The lively and dynamic public–private partnerships are capable of fully leveraging varied resources, capacities, and innovative means for effectively and comprehensively resolving the intricate problems of enabling greater AI access and promoting responsible digital netizenship. Through blending their combined experience and abilities, these stakeholders have the ability to make a huge impact toward closing the digital divide and making the digital society more inclusive and equitable.

7.4.2.5 Future Directions

Since technology is developing so fast and changing the nature of our modern world, policymakers and stakeholders must constantly reinvent and enhance their strategies to enable everyone to experience full and fair opportunities to participate in the numerous opportunities provided through AI. This requires a forward-looking and proactive stance entailing investment in future-facing technologies like 5G, satellite AI, and digital platforms and proactively and seriously grappling with emerging challenges such as but not limited to digital rights, the ubiquitous challenge of online disinformation, and the subterranean challenge of algorithmic bias. In doing so, the societies are actually able to unleash the disruptive potential of AI to produce sustainable and inclusive human development.

7.5 CONCLUSION

Overall, the theoretical framework of the use and access of digital AI as people's inherent right becomes a central discussion in modern digital society. By our examination of minority opinions and evidence-based facts, it's evident that the AI has an important role in the exercise and enjoyment of other intrinsic rights like freedom of expression, access to information, education, health, and political participation. Our discussion highlights the necessity of acknowledging AI access and usage as a key driver toward achieving human dignity and equality in the digital economy. The AI propels socioeconomic growth, enabling people and communities to close gaps, create inclusiveness, and improve the opportunities for advancement. By establishing AI access and use as a universal human right, policymakers and stakeholders will be in a position to respond more effectively to differences in digital access and provide everyone with equal chances to participate in the digital environment.

In addition, our analysis emphasizes the necessity of proactive measures toward bridging AI access and usage barriers, most importantly, by marginalized and disadvantageous groups. Efforts in expanding infrastructure, digital literacy, and accessible usage are important toward bridging the digital divide and enabling digital inclusion. Besides, protecting online freedom, privacy, and security is the utmost priority in maintaining the integrity of AI access as a human right, necessitating robust legal frameworks and enforcement mechanisms. Governments, international organizations, civil society, and the private sector must join forces in advancing an integrated approach to provide universal access to AI in future planning. Through accepting the right to access and utilization of AI as a human right, we are able to build a more equitable, inclusive, and participatory digital world in which everyone has the ability to harness AI's potential for good for humanity's benefit.

REFERENCES

Amodu, L., Omojola, O., Okorie, N., Adeyeye, B., & Adesina, E. (2019). Potentials of internet of things for effective public relations activities: Are professionals ready? *Cogent Business & Management*, 6, 1683951. https://doi.org/10.1080/23311975.2019.1683951

Arora, K., Joshi, G. P., Ragab, M., Rawa, M., Milyani, A. H., Mansour, R. F., & Yang, E. (2022). Bilateral contract for load frequency and renewable energy sources using advanced controller. *Computers, Materials & Continua*, 73(2), 3165–3180. https://doi.org/10.32604/cmc.2022.026966.

Arora, K., Kumar, A., Kamboj, V. K., Prashar, D., Shrestha, B., & Joshi, G. P. (2021). Impact of renewable energy sources into multi-area multi-source load frequency control of interrelated power system. *Mathematics*, 9(2), 186. https://doi.org/10.3390/math9020186.

Arora, K., Tripathi, S. L., & Kumar, S. (2022). *Smart Electrical Grid System: Design Principle, Modernization and Techniques*. CRC Press (Taylor & Francis Group).

Büchi, M., Just, N., & Latzer, M. (2016). Modeling the second-level digital divide: A five-country study of social differences in Internet use. *New Media & Society*, 18(11), 2703–2722. https://doi.org/10.1177/1461444815604154

Chemmencheri, S. R. (2016). Social protection as a human right in South Asia. Indian *Journal of Human Development*, 10(2), 236–252. https://doi.org/10.1177/0973703016671844

Chen, C.-C. (2015). Assessing the activeness of online economic activity of Taiwan's Internet Users: An application of the super-efficiency data envelopment analysis model. *Social Indicators Research*, 122(2), 433–451. https://doi.org/10.1007/s11205-014-0690-y.

Choudrie, J., Grey, S., & Tsitsianis, N. (2010). Evaluating the digital divide: The Silver Surfer's perspective. Electronic Government: *An International Journal*, 7(2), 148–167. https://doi.org/10.1504/EG.2010.030925

Llorente-Barroso, C., Viñarás, M., & Sánchez, M. (2015). Internet and the elderly: Enhancing active ageing. Comunicar, 23(45), 29–36.

Manickam, P., Girija, M., Dutta, A. K., Ramesh Babu, P., Arora, K., Jeong, M. K., & Acharya, S. (2024). Empowering cybersecurity using enhanced rat swarm optimization with deep stack-based ensemble learning approach. *IEEE Access*. https://doi.org/10.1109/ACCESS.2024.3395328.

Mascheroni, G., & Ólafsson, K. (2016). The mobile Internet: Access, use, opportunities and divides among European children. *New Media & Society*, 18(8), 1657–1679. https://doi.org/10.1177/1461444814567986

Singh, P., Arora, K., Rathore, U. C., Joshi, G. P., & Cho, W. (2024). Comparative study of controllers in battery energy storage system integrated with doubly fed induction generator-based wind energy conversion system for power quality improvement. *Energy Reports*, 11(9), 4587–4600. https://doi.org/10.1016/j.egyr.2024.04.020.

Wang, L., & Liu, C. (2018). The mobile Internet underclass: Reality or hyperbole? *Journal of Information Science*, 44(5), 569–579. https://doi.org/10.1177/0165551517690083

Chapter 8

Industrial IoT

Transforming connectivity and response in the digital age

Rajesh Singh, Anita Gehlot, Fraiz Parveen, Mahim Raj Gupta, and Praveen Kumar Malik

8.1 INTRODUCTION TO IIoT

Technological advances continue to digitize the real world, increase smart devices, and connect technologies and inventions to the cyber world. This trend leads to Industry 4.0, which is the fourth level of the industrial revolution in a structure aimed at turbocharging interdependence, automation, autonomy, machine learning (ML), and real-time data collection. More than 30 billion devices connected will be online by 2020, according to current forecasts compared to today's 20 billion [1]. Integration of the Internet of Things (IoT) with the ever-increasing industrial systems is indeed an enormous impetus for an idea now popularly known as the Industrial Internet of Things or IIoT. IIoT is that which extends to intelligent gadgets that gather, analyze, and exchange real-time events in industrial systems. This is to achieve operational efficiency, improved productivity, and subdivided management of assets and processes through customizable products, smart monitoring, and predictive machinery maintenance [2]. IIoT is a fast-developing technology that links numerous sensors to the internet, improving our daily lives. It has an influence on many elements of society, including the workplace, individual, and organizational lives. With increased internet access, connection prices are falling and better Wi-Fi-capable equipment is being created. The notion of linking items to the internet is becoming a new guideline for the future, altering how we interact with the environment [3]. IIoT and Industry 5.0 are two extensively explored areas of study, gaining attention from both academic and industry sectors. Industry 4.0 focuses on automation and production, with IIoT providing efficiency, flexibility, communication, adaptability, individualization, and modularity. As Industry 4.0 advances, it is critical to give a complete overview of the various study topics, as the authors acknowledge the necessity for a thorough grasp of the quickly changing IIoT framework [4]. Industry 5.0 is a digital revolution that will have an impact on many industries, including manufacturing. IIoT is a technical innovation that improves productivity and economic effect by allowing components to communicate with one another globally. Due to its distinct characteristics, the industrial industry confronts obstacles

 DOI: 10.1201/9781003641582-8

in integrating IIoT [5]. The fourth industrial revolution, IIoT, boosts productivity, reliability, and performance of competition, emphasizing profitability. Research studies on IIoT-enabled technologies have revealed the importance of world industrial growth, applications, problems, and future orientations. IIoT applications deal with challenges such as poor supply chain management, threats of data privacy breach, inefficient cloud strategy, and cost containment. Fog computing reduces emissions as it lowers emission for parking, platform, gasoline, and CO_2 [6]. The production has been visually modified to become more intelligent and data-centered; smart production units are now sophisticated networked industrial systems for handling materials, components, equipment, tools, inventories, and logistics networks. IIoT is the network of devices, sensors, and equipment affording remote access, effective monitoring, and improvement in the areas of data collecting, analysis, and transmission, as shown in Figure 8.1. IIoT will be necessary in Industry 4.0 for transforming the cyber-physical systems and production processes by means of big data and analytics. IIoT devices and network protocols open a sizable security window because they are too easy to attack and the data they carry is priceless. Security technologies are capable of detecting and mitigating intrusions; however, heterogeneity and interoperability are far beyond their measure. Due to the paucity of precise data regarding the activities, connectivities, and attack behaviors of IIoT systems, developing universal security solutions is crucial. Collecting and

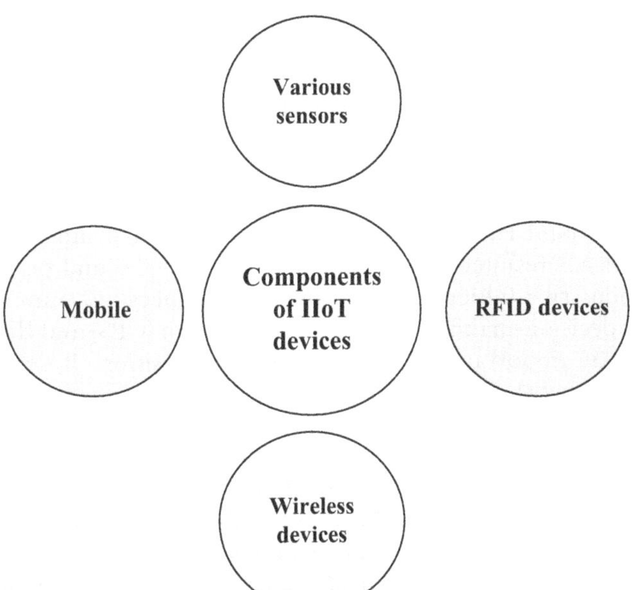

Figure 8.1 Components of IIoT devices.

producing such data is especially difficult owing to the multiplatform = communication protocols and multivendor devices [7].

IIoT is heavily contingent on data transmission for industrial output, particularly in the Industry 5.0 era. IoT devices may be linked to satellite networks to enable continuity and resilience during catastrophes. The data transmission window may be modified to fill the converged network's high bandwidth-delay product, and the threshold can be adjusted based on input to differentiate between distinct data losses during congestion avoidance. The heterogeneous network transmission model has been empirically validated [8]. IIoT is crucial for industrial sensors that monitor device status and machinery. However, dependability and security are critical considerations since these devices are susceptible to viruses, threats, and assaults. To safeguard millions of IIoT devices, an effective protection approach is required. Due to limited resources, IIoT at the cloud, fog, and edge has recently gained attention through research, providing the possibility to perform computationally intensive activities like security, data analytics, decision-making, and reporting through strong computing infrastructure [9].

8.2 INDUSTRIAL INTERNET OF THINGS 4.0 (IIoT4)

Industrial revolutions attempted to raise society's living standards and make life simpler for customers. Economic growth should coincide with each revolution, taking into account the expanding demand for power, non-renewable energy sources, and rising fuel prices. To achieve this need, the 21st century must work within these restrictions. Information and communication technologies (ICTs) are critical for developing creative solutions, with smart embedded systems and the internet playing key roles in driving ICT progress. By addressing these limits, the 21st century can improve living circumstances for the rising population [10]. Emerging technology and increased demand have prompted the industrial environment to swiftly evolve. This has resulted in a rise in processing devices and systems across sectors. Industry 4.0 focuses significantly on connection, communication, and smart decision-making. Cyber-physical system (CPS) and IIoT are projected to play important roles in the digital transition. Researchers and specialists are working on IIoT systems and CPS designs to link network devices from different ICT systems, virtualize assets, and integrate them with other industrial sectors [11]. The IoT has revolutionized the industrial sector by combining physical and digital aspects with minimal human participation. Its capacity to collect and analyze massive amounts of data, which is then employed by AI systems for activities such as performance optimization and maintenance, has resulted in substantial changes in the economy and business processes. However, the complexity of IIoT systems presents security and privacy issues, since they are vulnerable to sophisticated assaults and the loss of crucial, personal, and sensitive data [12]. The IoT and 5G advances have prompted interest in wireless solutions for crucial applications, notably IIoT. However, there has been little study on a unified design process that solves power, latency, and reliability complications, which is critical for future IIoT designs and development [13]. IoT enables data to be communicated via networks without requiring human intervention, boosting data transfer, and enhancing a wide range of applications such as agriculture, smart homes, healthcare, transportation, and the environment. It also improves industry productivity by enhancing monitoring and control, hence lowering expenses. The notion of IIoT, which makes use of radio-frequency identification (RFID), wireless, mobile, and sensor devices, presents a great chance to develop powerful industrial systems and applications. To better comprehend the evolution of IoT in many sectors, a number of IIoT applications have been developed [14]. The introduction of 5G-based IIoT in Industry 4.0 has expedited network automation while AI and human–robot interaction are progressing toward 6G-based Industry 5.0. However, further research is required to evaluate cutting-edge IIoT technologies in 5G-based Industry 4.0 and address issues for next-generation automation in 6G-based Industry 5.0, as shown in Table 8.1 [15]. This interconnectivity, proliferation of IIoT, smart ecosystems, and process automation have increased the potential but also enlarged the cyber-threat landscape. Convergence of industrial operational technology with information technology toward Industry 4.0 goals has brought sophistication, eco-complications of the IIoT ecosystem, paradigm shifts, and changes in architectures for security and digital forensic investigation [16]. IoT facilitates innovative business models in different industries through ubiquitous connectivity, effective data analytics, and decision support tools. However, due to the complexity of the IIoT, it is vulnerable to security breaches, and hence, the availability, confidentiality, and integrity cannot be guaranteed. This can lead to a lack of trust in network operations, loss of critical infrastructure, compromised safety, and privacy breaches on sensitive data [17].

Table 8.1 Industrial Internet of Things (IIoT) and its challenges

Topic	*Description*
IoT and IIoT applications [14]	IoT enhances efficiency across industries, while IIoT enables advanced industrial systems.
5G and 6G in IIoT [15]	5G accelerates IIoT automation, but research is needed for 6G-based Industry 5.0.
Cybersecurity challenges in IIoT [16]	IIoT expands cyber threats, complicating security and digital forensic investigations.
Security and privacy risks in IIoT [17]	IIoT's complexity increases security risks, affecting trust, safety, and data integrity.

8.3 KEY CHALLENGES OF IoT AND IIoT

IoT has transformed industrial processes by integrating devices, equipment, and systems, but it also raises data security concerns. IIoT trends and problems, as well as ideas for improving data security in IIoT ecosystems. Strong encryption, network segmentation, device management, anomaly detection, intrusion prevention systems, regulatory compliance, staff training, and incident response planning are some of the strategies used. Proper implementation assures authorized access, safeguards critical data, limits breaches, and reduces the risk of device compromise as described in Table 8.2 [18]. IoT is gaining popularity as it connects devices in industrial settings, providing several benefits but also posing security issues. Despite parallels to the consumer IoT, protecting the industrial IoT poses both difficulties and possibilities due to longer component lives and bigger network sizes. This trend is fueled by the success of IoT in networking consumer devices [19].

Understanding solution areas and integrating manufactured systems in supply chain management are necessary for IIoT design. It covers a wide range of industrial applications, including real-time monitoring, machine operability surveillance, inventory management, and predictive maintenance. Energy efficiency, performance, interoperability, security, and privacy are all challenges. Sensors and networks must be implemented in cyber-physical systems for remote industrial applications to function. IIoT must address access control, scalability, and functional security to assure authentication, data integrity, and anonymity in various contexts [20]. IIoT and Industry 5.0 are revolutionizing industrial processes by strengthening device and machine monitoring and control. IIoT improves manufacturing operations by increasing data accessibility, safety, and maintenance forecasts. Advanced

Table 8.2 Key aspects and challenges of IoT and IIoT

Topic	*Description*
IIoT technologies and applications [14]	IIoT uses RFID, mobile, sensor devices and wireless, and applications have been made to understand its impact on various sectors.
5G-based IIoT and Industry 4.0 [15]	The introduction of 5G-based IIoT has fastened network automation, while advancements such as human-robot and AI interaction have led toward Industry 5.0 and 6G.
IIoT and security challenges [17]	IIoT systems are vulnerable to security breaches, making it difficult to ensure data availability, confidentiality, and integrity, leading to potential privacy breaches and safety risks.
Strategies for securing IIoT [18]	Approaches include strong encryption, network segmentation, device management, anomaly detection, intrusion prevention systems, and staff training.
Challenges in protecting industrial IoT [19]	Protecting industrial IoT poses unique challenges due to longer component lifespans and larger network sizes compared to consumer IoT systems.

technologies provide operational efficiency, connectivity, scalability, and interoperability, as well as time and cost savings. To provide latency-free data transfer, services must have rapid reaction times and be integrated with high-speed cloud platforms [21]. IoT has allowed smart machine–machine interactions that have transformed human–machine communication and decision-making. Over the last 20 years, these systems have grown in importance in the commercial and industrial sectors. IIoT, which consists of engineering equipment, may enhance industrial operations. However, rising energy use and network traffic may complicate this endeavor [22].

8.4 UTILIZATION OF IIoT IN MANUFACTURING

The fast proliferation of IoT devices, particularly in industrial and manufacturing settings, has prompted more study into energy efficiency in green IIoT (G-IIoT). Limited resources can have a huge influence on the lifespan of these systems, as big sensors and devices consume more energy, increasing the carbon footprint and decreasing their overall sustainability [23]. With real-time data gathering and analysis, IIoT allows automation and efficiency through better decision-making that is based on data. With traditional architectures, it becomes complicated to deal with huge data volumes, bandwidth constraints at the edges, and the privacy of data. Distributed edge-to-cloud computing offers the opportunity of edge-to-cloud data analytics along with distributed AI for enhanced decision-making capabilities and predictive maintenance. However, detailed investigations are needed to establish essential needs and assess the extent to which developing IoT systems fit these objectives [24].

Digitalization of production processes, higher connectivity, and automation in smart factories and industrial systems have been made a reality by the IIoT revolution in Industry 4.0. System dependability is one of the essentials for performance but achieving reliability in IIoT systems is an issue due to variable and dynamic industry contexts as well as complexity in industrial systems [25]. The massive amounts of data produced by the IoT require more intelligent processing techniques such as deep learning (DL). The applicability of IIoT on DL can add new applications of smart assembly and manufacturing, effective networking, accident detection, and prevention [26]. Figure 8.2 shows the utilization of IIoT in manufacturing.

8.5 CASE STUDIES

8.5.1 Tom farms

Farming is the leading industry in implementing IoT for industrial use cases. One well-known case study that focuses on the usage of IoT in industrial farming is Bytes and Bushels. Tom Farms, a family-run farm in the

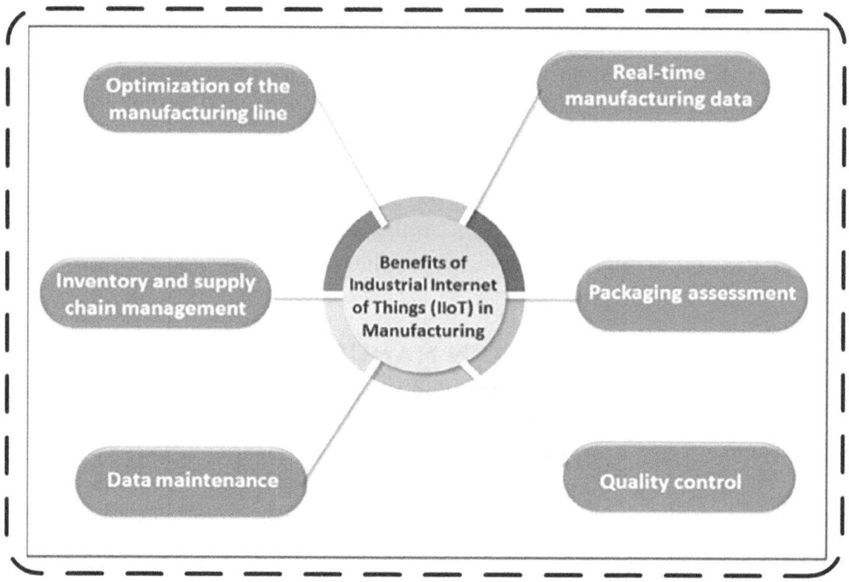

Figure 8.2 Utilization of IIoT in manufacturing.

United States, is the subject of the IoT case study on industrial automation in agriculture. Tom Farms has figured out a creative way to use IoT to make it easier to distribute its products. Tom Farms concentrated on the development of new production technologies and acquired additional land after the farm crisis of the 1980s. Tom Farms successfully implemented sensors, GPS information, self-driving tractors, and iPhone apps for effective irrigation.

8.5.2 The Airbus

Airbus would come in at number two when it comes to instances of industrial IoT protocols. To increase manufacturing capacity and streamline operations, the commercial airplane maker is utilizing the IoT to build a future of the factory. To monitor every stage of the assembly process and prevent errors, Airbus uses sensors that are built into various tools and machinery. Furthermore, Airbus has implemented wearable technologies for employees, such as industrial smart glasses, which may enhance workplace security.

8.5.3 The Internet of Things by Tesla vehicle

AI and IoT play a key role in Tesla Motors' goal for self-driving automobiles. IoT-based solutions can be used by the futuristic auto manufacturer to customize vehicles to meet the needs of its customers. The IoT supports key components of Tesla's self-driving cars, including connectivity, autopilot mode, and mobile application control.

8.5.4 Caterpillar

Moreover, the collection of industrial IoT application cases illuminates instances such as Caterpillar. The manufacturer of heavy machinery has been a prominent pioneer in the deployment of the IIoT. Caterpillar, commonly referred to as CAT, has benefited from the application of IoT in industrial settings.

Machine operators may now access real-time information on fuel levels and machine health status thanks to the heavy-equipment manufacturer's use of IoT and augmented reality technologies. To further protect assets on boats and transport vessels, CAT's maritime asset intelligence division has implemented sensor-driven analytics.

8.5.5 The Hitachi

Hitachi is another well-known company that has been included in IoT case studies on industry automation. A notable feature of Hitachi's IIoT use cases is its committed team of 16,000 IoT-focused professionals. The business offers numerous IoT-based solutions in addition to Lumada, an IoT platform. Additionally, Hitachi has been developing an IoT-based production model that may shorten lead times for infrastructure manufacturing activities.

8.5.6 Deere John

John Deere, the well-known manufacturer of tractors. The company uses the IoT to power its upcoming line of self-driving tractors. Above all, John Deere's proficiency with GPS technology helps its efforts to create self-driving tractors. Additionally, telematics is used by John Deere to facilitate and forecast the advantages of maintenance.

8.5.7 The Amazon

The last addition to the industrial IoT case studies highlights Amazon, the retail behemoth. It is one of the pioneers in utilizing the IoT to revolutionize warehousing and logistics processes. For instance, linked robots are used in Amazon's fulfillment centers to find product shelves and deliver them to employees. In addition to increasing operational efficiency, the deployment of IoT-connected robots may enable considerable cost savings.

8.5.8 The Intel

With its partner NEXCOM, Intel has ventured into the IIoT space to revolutionize electronic medical record administration. NEXCOM's innovative IoT-based medical informatics system aids in removing obstacles to medical device and IoT network connection. Data from medical devices can be more easily documented in electronic medical records and sent to healthcare providers' private clouds.

8.6 DISCUSSION

- IIoT technology utilizes sensors, data analysis, and cameras to provide insights into innovations in industry, enabling devices to predict difficulties as they advance and businesses to make more informed, faster choices in the future.
- As far as benefits relate to industrial enterprises, IoT offers increased levels of efficiency, productivity, and cost savings, but it also poses huge privacy threats that must be addressed in order to achieve network and end-user reliability.
- Privacy by design principles may ensure that privacy is built in from the beginning while designing IIoT systems and devices.
- Despite its limitations, IIoT infrastructure is critical to the success of immense businesses, and without it, the current industrial revolution would be impossible to operate efficiently.
- Seeking vibrant and completely vibrant workplaces, the Industry 5.0 postdisaster communication uses heterogeneous Network Satellite and Multihop networks, keeping production continuity in the intelligent production systems even during infrastructure failure.
- The potential for IoT in industrial applications including automated monitoring and control has generated interest in its implementation. However, the growing number of devices may use substantial energy, posing a bottleneck in blocking its proliferation.
- The information obtained from network traffic, host resources, logs, and security mechanism warnings pertains to new IIoT connection protocols and recent attack strategies. An exploration tool is utilized for exploring it, as well as ML and DL approaches.

8.7 CONCLUSION

This move will be used to encourage more amendments to improve the practice of the industry. Production efficiency is increased not only through greater automation but also through a radical change in decision-making from nonempirical to data-driven decisions. Among the fast-growing topics, promising an entirely new revolutionary perspective on the operation and administration of IoT-based applications is a distributed edge-to-cloud computing. In this paper, we discuss the contribution of distributed edge-to-cloud computing in IIoT, key requirements for it, and capabilities comparison of three major IoT systems with regard to performance, energy consumption, scalability, interoperability, fault tolerance, data analytics capabilities, security features, and deployment options. Thus, IIoT emerges as one of the major domains in the newly initiated revolution by Industry 4.0. The article summarizes issues and opportunities concerning the traditional technology and applications of IIoT for industrial growth. In so doing, it brings different examples from the application of other sectors and recaps past research conducted related to them for its audience to become aware of what has been achieved so far in that particular domain. Developments in the production process, supply chain, robotic facilities, embedded systems, and networked equipment all have an impact on real-time manufacturing efficiency. This leads to reduced risk and more creativity. Automated industries are becoming more productive because of low-cost, reliable sensors. IIoT technology may be utilized to self-diagnose, repair equipment, and assemble lines, lowering downtime, costs, worker productivity, and energy consumption. This technology also enables firms to reduce operating costs and generate new revenue streams. Fast and effective production and supply chain operations reduce product cycle time. This presents a discussion on the security requirements and approaches intended for future IIoT in the context of Industry 4.0. Thereby demonstrating that even as compared to the consumer IoT, IIoT has its unique security objectives and concerns. The research suggests that various counteractions are needed, with some being built primarily for consumer contexts and adopted for industrial use.

REFERENCES

1. Munirathinam, S. (2020). Industry 4.0: Industrial internet of things (IIOT). In: A. R. Hurson (Ed.), *Advances in Computers* (Vol. 117, pp. 129–164). Elsevier.
2. Khan, W. Z., Rehman, M. H., Zangoti, H. M., Afzal, M. K., Armi, N., & Salah, K. (2020). Industrial internet of things: Recent advances, enabling technologies and open challenges. *Computers & Electrical Engineering*, *81*, 106522.
3. Malik, P. K., Sharma, R., Singh, R., Gehlot, A., Satapathy, S. C., Alnumay, W. S., ... & Nayak, J. (2021). Industrial Internet of Things and its applications in industry 4.0: State of the art. *Computer Communications*, *166*, 125–139.
4. Bhatt, B., Sharma, H., Arora, K., Joshi, G. P., & Shrestha, B. (2023). Levy flight-based improved grey wolf optimization: A solution for various engineering problems. *Mathematics*, *11*(7), 1745. https://doi.org/10.3390/math11071745.
5. Alabadi, M., Habbal, A., & Wei, X. (2022). Industrial internet of things: Requirements, architecture, challenges, and future research directions. *IEEE Access*, *10*, 66374–66400.

6. Ahmed, S. F., Alam, M. S. B., Hoque, M., Lameesa, A., Afrin, S., Farah, T., ... & Muyeen, S. M. (2023). Industrial Internet of Things enabled technologies, challenges, and future directions. *Computers and Electrical Engineering*, *110*, 108847.
7. Al-Hawawreh, M., Sitnikova, E., & Aboutorab, N. (2021). X-IIoTID: A connectivity-agnostic and device-agnostic intrusion data set for industrial Internet of Things. *IEEE Internet of Things Journal*, *9*(5), 3962–3977.
8. Zong, L., Memon, F. H., Li, X., Wang, H., & Dev, K. (2021). End-to-end transmission control for cross-regional industrial Internet of Things in Industry 5.0. *IEEE Transactions on Industrial Informatics*, *18*(6), 4215–4223.
9. Abuhasel, K. A., & Khan, M. A. (2020). A secure industrial internet of things (IIoT) framework for resource management in smart manufacturing. *IEEE Access*, *8*, 117354–117364.
10. Tabaa, M., Monteiro, F., Bensag, H., & Dandache, A. (2020). Green Industrial Internet of Things from a smart industry perspectives. *Energy Reports*, *6*, 430–446.
11. Pivoto, D. G., De Almeida, L. F., da Rosa Righi, R., Rodrigues, J. J., Lugli, A. B., & Alberti, A. M. (2021). Cyber-physical systems architectures for industrial internet of things applications in Industry 4.0: A literature review. *Journal of Manufacturing Systems*, *58*, 176–192.
12. Demertzi, V., Demertzis, S., & Demertzis, K. (2023). An overview of privacy dimensions on the industrial internet of things (IIoT). *Algorithms*, *16*(8), 378.
13. Da Costa, C. M., & Baltus, P. (2020). Design methodology for industrial Internet-of-Things wireless systems. *IEEE Sensors Journal*, *21*(4), 5529–5542.
14. Ahmed, H., Ramadan, A. A., Elkordy, E. H., & Elngar, A. A. (2022). Introduction to Industrial Internet of Things (IIoT). In: A. A. Elngar, A. A. Ramadan, and H. Ahmed (Eds.), *Industrial Internet of Things* (pp. 1–18). CRC Press.
15. Chi, H. R., Wu, C. K., Huang, N. F., Tsang, K. F., & Radwan, A. (2022). A survey of network automation for industrial internet-of-things toward industry 5.0. *IEEE Transactions on Industrial Informatics*, *19*(2), 2065–2077.
16. Kebande, V. R. (2022). Industrial internet of things (IIoT) forensics: The forgotten concept in the race towards industry 4.0. *Forensic Science International: Reports*, *5*, 100257.
17. Arora, K., Kumar, A., Kamboj, V. K., Prashar, D., Jha, S., Shrestha, B., & Joshi, G. P. (2020). Optimization methodologies and testing on standard benchmark functions of load frequency control for interconnected multi-area power system in smart grids. *Mathematics*, *8*(6), 980. https://doi.org/10.3390/math8060980.
18. Bhatia, D. (2024). Data security for industrial internet of things: An essential perspective for industrial engineering. In: A. K. Tyagi, A. K. Solanki, and P. Tyagi (Eds.), *Industrial Internet of Things Security* (pp. 71–80). CRC Press.
19. Serror, M., Hack, S., Henze, M., Schuba, M., & Wehrle, K. (2020). Challenges and opportunities in securing the industrial internet of things. *IEEE Transactions on Industrial Informatics*, *17*(5), 2985–2996.
20. Kantanavar, P., & Rajendran, S. (2023). Industrial internet of things: Architecture and its applications. In: R. M. Kannan and A. M. Madan (Eds.), *Internet of Things* (pp. 237–256). Chapman and Hall/CRC.

21. Balakera, N., Konstantinidis, F. K., Tsimiklis, G., Latsa, E., & Amditis, A. (2023, February). Iiot network system from data collection to cyber-physical system transmission under the industry 5.0 era. In: *International Congress on Information and Communication Technology*, London, United Kingdom (pp. 929–941). Springer Nature, Singapore.
22. Arora, K., Kumar, A., Kamboj, V. K., Prashar, D., Shrestha, B., & Joshi, G. P. (2021). Impact of renewable energy sources into multi-area multi-source load frequency control of interrelated power system. *Mathematics*, *9*(2), 186. https://doi.org/10.3390/math9020186.
23. Mao, W., Zhao, Z., Chang, Z., Min, G., & Gao, W. (2021). Energy-efficient industrial Internet of things: Overview and open issues. *IEEE Transactions on Industrial Informatics*, *17*(11), 7225–7237.
24. Arora, K., Joshi, G. P., Ragab, M., Rawa, M., Milyani, A. H., Mansour, R. F., & Yang, E. (2022). Bilateral contract for load frequency and renewable energy sources using advanced controller. *Computers, Materials & Continua*, *73*(2), 3165–3180. https://doi.org/10.32604/cmc.2022.026966.
25. Kim, D. S., Hoa, T. D., & Thien, H. T. (2022). On the reliability of industrial internet of things from systematic perspectives: Evaluation approaches, challenges, and open issues. *IETE Technical Review*, *39*(6), 1277–1308.
26. Khalil, R. A., Saeed, N., Masood, M., Fard, Y. M., Alouini, M. S., & Al-Naffouri, T. Y. (2021). Deep learning in the industrial internet of things: Potentials, challenges, and emerging applications. *IEEE Internet of Things Journal*, *8*(14), 11016–11040.

Chapter 9

Harnessing data analytics for intelligent electrical machine design

Satya Ravi Teja Peddada, Ravi Shankar Saripalle, Jayasri Angara, and Praseeda Saripalle

9.1 INTRODUCTION

Electrical machine design is on the cusp of a paradigm shift. Historically based on analytical modeling, iterative prototyping, and simulation-based optimization, it is now increasingly influenced by the advent of data-centric engineering. This is motivated by the evolution of Industry 4.0 approaches focused on digitization, automation, and real-time analysis within engineering activities (Kagermann et al., 2013). In particular, the extensive usage of sensor-enabled machines, digital twins, and cloud-based simulation environments has facilitated the generation of vast amounts of performance and design information across a wide variety of machine types and operating conditions (Fuller et al., 2020; Hu et al., 2021).

Electrical drives such as induction motors, brushless DC (BLDC) motors, permanent magnet synchronous machines (PMSMs), and switched reluctance motors (SRMs) are the backbone of modern electromechanical systems. They are essential in electric transportation, renewable energy systems, high-end manufacturing, robotics, and aerospace technology. Their performance is controlled by delicate interactions of electromagnetic geometry, thermal conductivity, material properties, and control logic. Previously, reducing these qualities has involved hundreds or thousands of expensive simulations utilizing finite element methods (FEM) or multi-domain modeling software such as Simulink, Motor-CAD, or COMSOL Multiphysics (Sudhoff, 2021; MathWorks, 2023; ANSYS Inc., 2023; Motor Design Ltd., 2023; COMSOL Inc., 2023). In this chapter, we are working with a mid-size BLDC motor topology popular with e-bikes, mini electric vehicles, and industrial automation controls, wherein torque values are in the range of roughly 5–25 Nm.

Although such tools deliver high-fidelity results, they have a tendency to consume a lot of computational horse power, slow down design iteration loops, and offer limited system-level design understanding. This is where data analytics becomes transformative. With access to structured data, engineers can apply statistical modeling, dimensionality reduction,

 DOI: 10.1201/9781003641582-9

regression analysis, clustering, and machine learning techniques to make decisions that are faster, more informed, and adaptive (Bhatt et al., 2023; Goodfellow et al., 2016; Bishop, 2006).

For example, using historical simulation or test datasets, engineers can train surrogate models to estimate torque, losses, or efficiency directly from input features eliminating the need to repeatedly run full simulations. Unsupervised learning can reveal performance clusters or identify high-potential design regions. This workflow significantly reduces computational expense and speeds up early-stage prototyping.

Notably, this shift has already been demonstrated in real-world thermal estimation of power electronics and drives (Peddada et al., 2019) and continues to expand via digital twin platforms that integrate sensing, simulation, and analytics in real time (Fuller et al., 2020; Sharma et al., 2020).

This chapter introduces the foundational concepts, methods, and MATLAB-based workflows for applying data analytics to electric machine design. Through real or synthetic datasets, readers are guided step-by-step through

- Data acquisition, preprocessing
- Exploratory data analysis (EDA)
- Regression modeling and prediction
- Dimensionality reduction and visualization
- Feature selection and engineering
- A complete case study on torque prediction for a BLDC motor

By the end of this chapter, readers will be in a position to incorporate data-driven design into their design workflows, accelerating innovation while maintaining engineering rigor.

9.2 DATA SOURCES IN ELECTRICAL MACHINE DESIGN

Application of data analytics in electric machine design starts with understanding sources and data structure. Data are typically classified into three broad categories.

9.2.1 Simulation-based data

Simulation data is generated using finite element method (FEM) software such as ANSYS Maxwell, COMSOL Multiphysics, or Motor-CAD. These software packages simulate the electromagnetic, thermal, and mechanical properties of electrical machines across various operating conditions.

Major output parameters include

- Magnetic flux densities across different regions
- Core losses and copper losses
- Torque and torque ripple over time
- Efficiency maps across speed–torque operating ranges
- Temperature profiles across windings and stator cores

Simulation results are typically structured as multidimensional arrays or high-dimensional tables, where each row represents a design configuration, and each column corresponds to an output variable. Large-scale parametric sweeps can easily produce gigabytes of data, making this source best for building surrogate models as well as training machine learning algorithms for design space exploration.

9.2.2 Experimental data

Experimental results are obtained by physical testing of prototypes or completed machine assemblies. This data provides ground-truth validation of simulation models and captures machine behavior under real-world loading conditions.

Typical variables measured include

- Input voltage and current waveforms
- Output torque and rotational speed
- Stator and rotor temperatures
- Vibrations, acoustic emissions, and mechanical stress

Commonly Used Sensors:

- Hall effect current sensors
- Thermocouples and RTDs (Resistance Temperature Detectors)
- Optical and magnetic encoders
- Piezoelectric accelerometers

Experimental setups often use software platforms such as NI LabVIEW, the MATLAB Data Acquisition Toolbox, or dedicated PLC/SCADA systems. Data is collected in formats like .CSV, .TDMS, or .MAT and may involve hundreds of channels of time-series data. Preprocessing steps such as filtering, synchronization, and interpolation are often required before analytics.

9.2.3 Operational field data

Operational field data is gathered from deployed machines in real environments, such as electric vehicles, wind turbines, or manufacturing plants.

These measurements are recorded using IoT-enabled systems, edge computing devices, or on-board loggers and often span days or months.

Characteristics of Operational Data:

- Continuous collection over long durations
- High-frequency sampling (1 kHz–100 kHz)
- Associated metadata (timestamp, GPS, ambient temperature)
- Cloud-based storage and wireless transmission

Examples of Field Measurements:

- Real-time current, voltage, and power consumption
- Thermal behavior across critical components
- Uptime statistics, usage patterns, and error/fault logs
- Closed-loop control signals and diagnostics.

This rich and voluminous dataset is especially valuable for predictive maintenance, condition monitoring, and anomaly detection using machine learning models. It enables engineers to identify complex interactions among operational variables, detect inefficiencies early, and develop robust performance prediction pipelines (Table 9.1).

Example MATLAB Code:

```
data = readtable('Motor_Design_Data.csv');
head(data)
```

Case Example:

- A test bench conducts **100 motor experiments** by systematically varying
  - Input voltage (180–240 V)
  - Load current (5–15 A)
  - Rotor speed (1,000–4,000 RPM)

Table 9.1 Sensor signals for electrical motor characterization

Signal name	*Unit*	*Sampling rate*	*Typical source*
Voltage	V	1 kHz	Power supply
Current	A	1 kHz	Hall effect sensor
Speed	RPM	10 kHz	Incremental encoder
Ambient temperature	°C	1 Hz	Thermocouple
Cooling airflow	CFM	10 Hz	Flow sensor/fan control system
Torque	Nm	1 kHz	Torque transducer

- Ambient temperature (20°C–45°C)
- Cooling airflow (50–200 CFM)

For Each Experiment, the System Logs:

- Mechanical Torque Output (Nm)
- All corresponding input and environmental conditions

This structured dataset becomes a foundational input for statistical modeling and machine learning workflows in the design process. In MATLAB, these signals can be easily imported using `readtable`, `readmatrix`, or interfaces to real-time systems via Simulink. The data is typically stored in `.CSV`, `.MAT`, or `.TDMS` formats and is structured as a tabular matrix with 100 rows and six columns.

Such Datasets are Ideal for

- Building predictive models for torque output
- Performing exploratory data analysis (EDA) using correlation plots and PCA
- Identifying design space clusters and performance regimes

By applying analytics tools to this dataset, engineers can significantly accelerate early-stage design decisions, reduce testing overhead, and derive data-backed insights for robust electric motor optimization.

9.3 DATA PREPROCESSING TECHNIQUES

Raw data is rarely ready for analysis as there could some information missing or some corrupt (very noisy signals). Preprocessing is a crucial step that ensures the quality, consistency, and interpretability of the input data for downstream analytical tasks. Poorly preprocessed data can lead to misleading results, model instability, and loss of actionable insights. This section outlines essential preprocessing tasks with relevant techniques and MATLAB implementations. Key preprocessing tasks include:

9.3.1 Handling missing and corrupted data

In engineering datasets, we often see issues caused by sensor malfunctions, communication lags, environmental interference, or manual recording errors. Such data losses are known as missing data. The strategies for addressing missing values depend on the context and extent of the data loss:

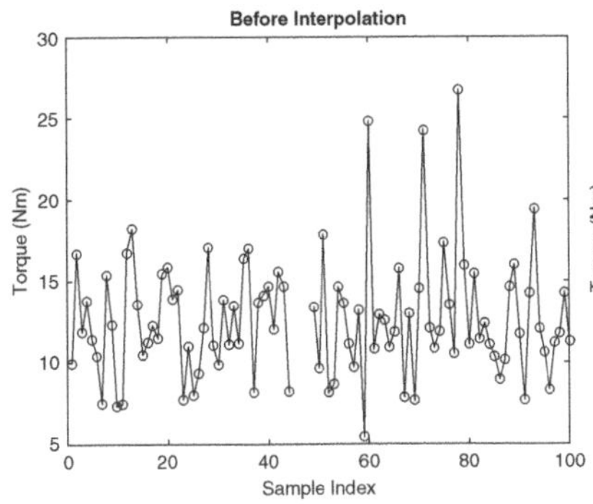

Figure 9.1 Interpolation of missing torque values between 45 and 48 samples using a linear method.

- **Interpolation:** Suitable for continuous time-series data. Techniques include linear interpolation, spline interpolation, and piecewise cubic Hermite interpolating polynomials (PCHIP).
- **Imputation with Central Tendency:** Replace missing entries with mean, median, or mode values, especially effective for normally distributed data.
- **Row/column Deletion:** Appropriate when missing values are scarce and randomly distributed.
- **Forward/Backward Fill:** Useful in sequential data, where previous or next known values are propagated (Figure 9.1).

MATLAB Example:

```
% Simulate missing values in Torque (for demonstration)
data = readtable('Motor_Design_Data.csv');
data.Torque(45:48) = NaN;  % Introduce missing values

% Plot before interpolation
figure;
plot(data.Torque); title('Before Interpolation');

% Linear interpolation
data.Torque = fillmissing(data.Torque, 'linear');

% Plot after interpolation
figure;
```

9.3.2 Normalization and scaling

Feature scaling is essential before applying algorithms that rely on distance computations (e.g., K-means and PCA) or gradient descent optimization (e.g., regression models).

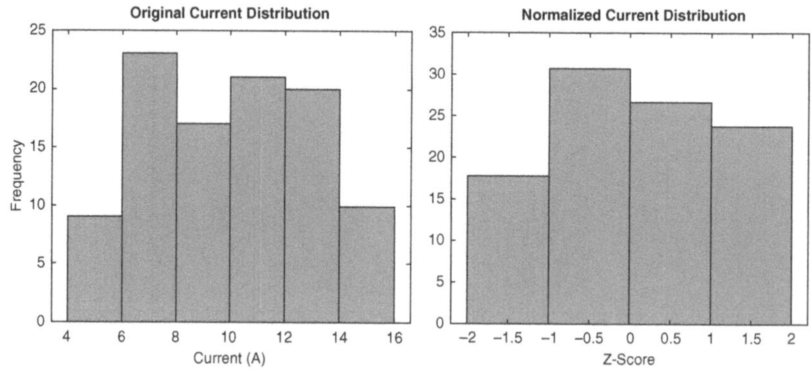

Figure 9.2 Effect of Z-score normalization on the distribution of current.

Common Scaling Techniques:

- **Min-Max Scaling:** Transforms values to a fixed range [0, 1].
- **Z-score Normalization:** Centers data around zero with a unit standard deviation.
- **Log Transformation:** Helps normalize right-skewed variables (Figure 9.2).

MATLAB Example:

```
% Z-score normalization for 'Current'
figure;
histogram(data.Current); title('Original Current
Distribution');
data.Current = (data.Current - mean(data.Current)) / std(
data.Current);

figure;
histogram(data.Current); title('Normalized Current
Distribution');
```

9.3.3 Outlier detection

Outliers are data points that deviate significantly from the overall pattern. They can represent real extreme events (e.g., overloads) or noise. Identifying and handling them appropriately improves model robustness.

Detection Techniques:

- **Boxplots:** Visual inspection using interquartile ranges (IQR).
- **Z-score Thresholding:** Flags values exceeding a set standard deviation (typically ±3).
- **Isolation Forests:** Anomaly detection algorithm for high-dimensional datasets (Figure 9.3).

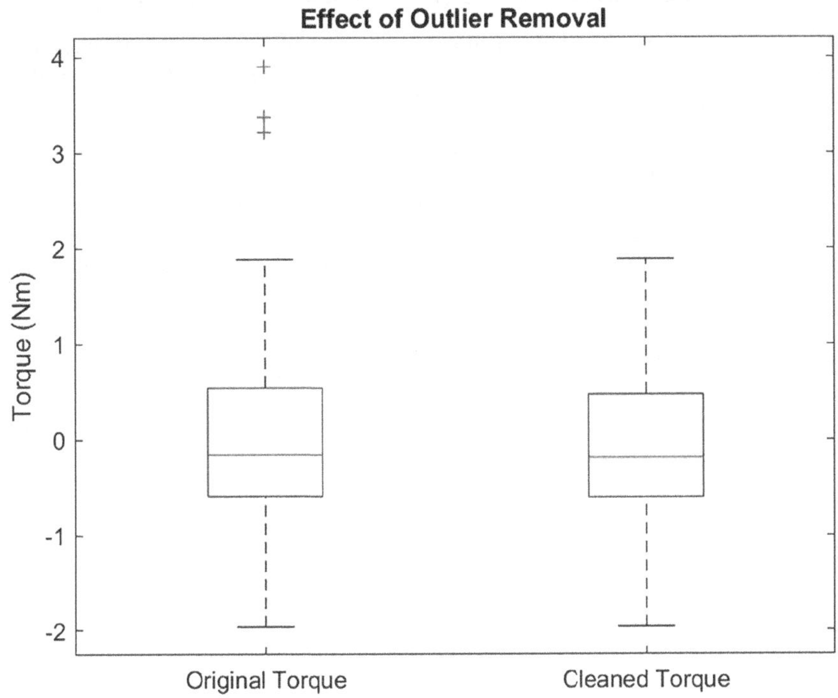

Figure 9.3 Comparison of torque values before and after outlier values removal.

MATLAB Example:

```
% Normalize the full dataset for distance-based outlier
analysis
data_scaled = normalize(data);

% Z-score based outlier detection in Torque
z = zscore(data_scaled.Torque);
outliers = abs(z) > 3;
cleaned_data = data_scaled(~outliers);

fprintf('Removed %d outliers from Torque\n', sum(outliers));

% Boxplot visualization
figure;
boxplot([data_scaled.Torque, cleaned_data.Torque], ...
    'Labels', {'Original Torque', 'Cleaned Torque'});
title('Effect of Outlier Removal');
ylabel('Torque (Nm)');
```

9.3.4 Categorical encoding (for mixed-type datasets)

While not always needed in numerical datasets, categorical encoding becomes relevant when including motor type, control mode, or manufacturer as input features.

- **One-Hot Encoding:** Converts categories to binary vectors.
- **Label Encoding:** Assigns integer codes to categories (not suitable for unordered data).
- **MATLAB Tip:** Use `dummyvar` or `categorical` functions to manage such variables.

9.3.5 Feature engineering considerations

Preprocessing is often combined with feature engineering to extract meaningful insights. Examples include:

- Creating derived variables (e.g., power=voltage×current)
- Aggregating rolling means for time-series smoothing
- Constructing domain-specific ratios or indices (e.g., torque-to-weight ratio)

These steps ensure that the dataset is high-quality, model-ready, and optimized for extracting meaningful patterns in later stages of analysis.

9.4 EXPLORATORY DATA ANALYSIS

Exploratory Data Analysis (EDA) is a crucial step in any data analytics pipeline. It involves visually and statistically examining datasets to uncover underlying patterns, identify anomalies, assess distributions, and formulate hypotheses. EDA informs feature selection, model choice, and data transformation strategies, laying the foundation for reliable statistical modeling and machine learning.

9.4.1 Correlation analysis

Correlation analysis measures the strength and direction of linear relationships between pairs of variables. Highly correlated input features may introduce redundancy or multicollinearity into predictive models.

MATLAB Example:

```
figure;
corrplot(data_scaled,'varNames', data_scaled.Properties.
VariableNames);
title('Correlation Matrix of Normalized Features');
```

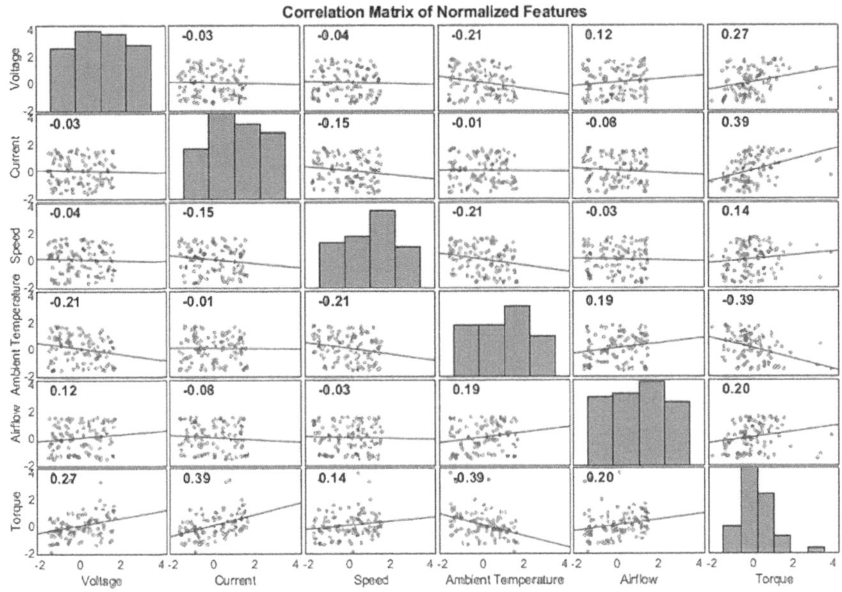

Figure 9.4 Correlation matrix and scatter plots for selected motor design parameters.

This command produces a matrix of scatter plots and Pearson correlation coefficients between all variables in the dataset (Figure 9.4).

Interpretation:

- A strong positive correlation ($r > 0.8$) between `Voltage` and `Torque` may suggest proportional loading.
- A strong negative correlation between `Speed` and `Current` could imply back-EMF effects.

9.4.2 Principal component analysis (PCA)

PCA is a powerful dimensionality reduction technique that transforms correlated variables into a new set of uncorrelated variables (principal components), ordered by variance explained.

Use Cases in Motor Design:

- Visualizing design clusters in 2D/3D
- Removing redundant dimensions
- Highlighting dominant parameter combinations (Figure 9.5)

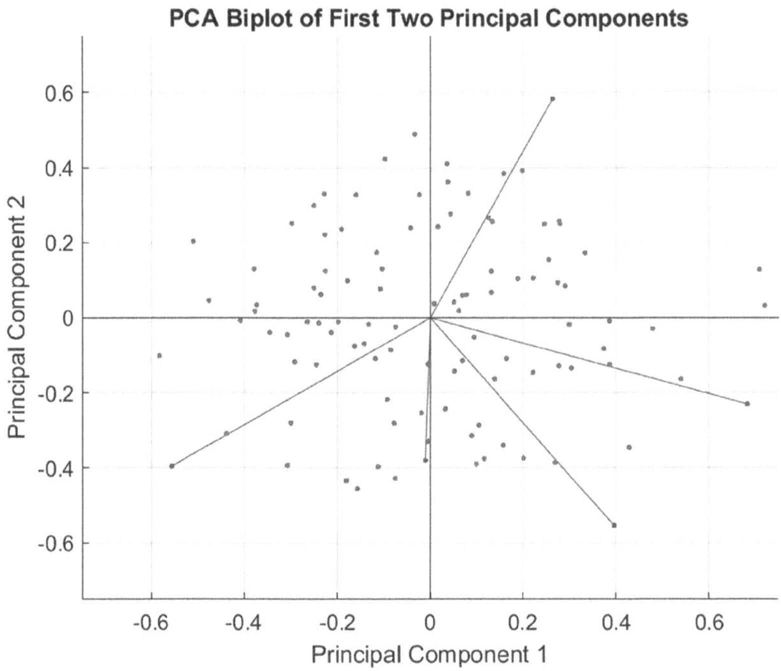

Figure 9.5 PCA biplot showing projections of original variables and samples onto the first two principal components.

MATLAB Example:

```
[coeff, score, latent] =
pca(table2array(data_scaled(:,2:end)));
figure;
biplot(coeff(:,1:2), 'Scores', score(:,1:2));
title('PCA Biplot of First Two Principal Components');
xlabel('Principal Component 1'); ylabel('Principal Component
2');
```

Interpretation:

- PC1 may represent overall electrical load
- PC2 could reflect thermal behavior separation (e.g., airflow vs. temperature)

9.4.3 Heatmaps and scatter matrices

Heatmaps and scatter matrix plots provide compact visual representations of pairwise interactions and variable distributions (Figure 9.6).

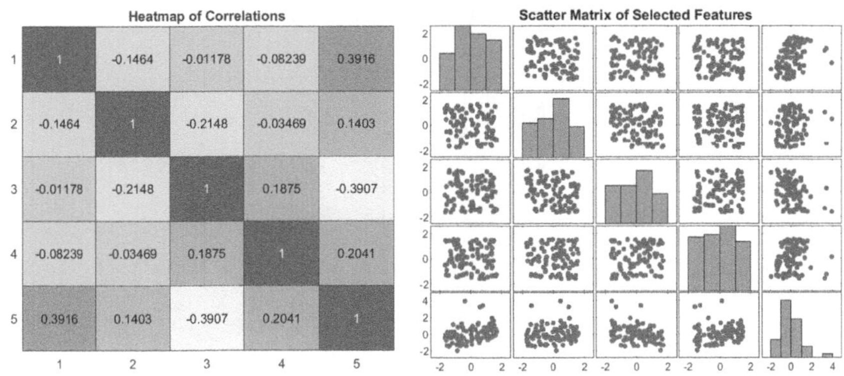

Figure 9.6 Correlation heatmap revealing clusters of closely related parameters (e.g., voltage, current, and torque) and scatter matrix illustrating bivariate patterns and linear/nonlinear trends.

MATLAB Example:

```
figure;
heatmap(corrcoef(table2array(data_scaled(:,2:end))));
title('Heatmap of Correlations');
figure;
plotmatrix(table2array(data_scaled(:,2:6)));
title('Scatter Matrix of Selected Features');
```

9.4.4 Distribution analysis

Understanding the distribution of each variable is essential for

- Choosing appropriate statistical tests
- Detecting skewness, kurtosis, and multimodal behaviors
- Evaluating normality assumptions (critical for linear models)

Common Plots:

- Histograms
- Kernel Density Estimations (KDE)
- Q-Q plots (quantile-quantile) (Figure 9.7)

MATLAB Example:

```
figure;
histogram(data.Torque);
title('Histogram of Torque Values');
[f, xi] = ksdensity(data.Torque);
figure;
```

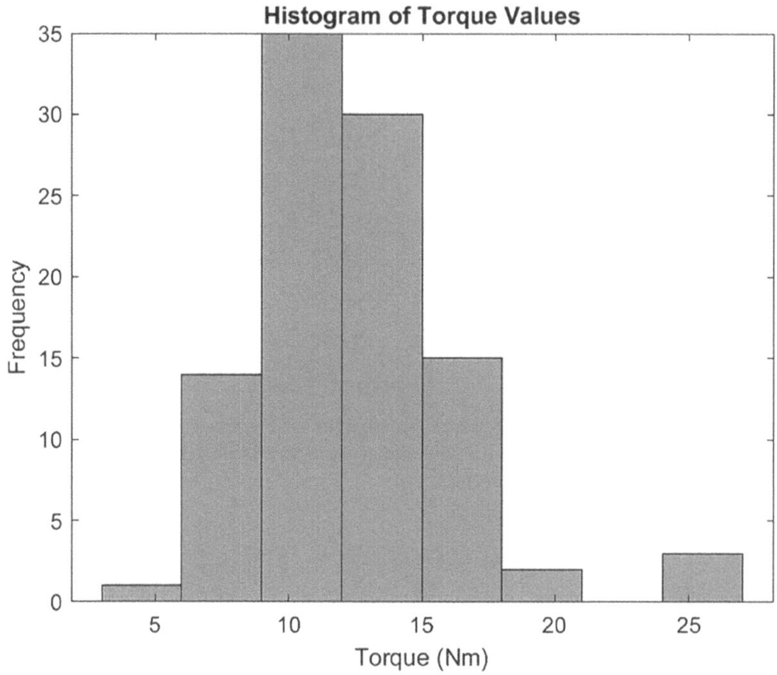

Figure 9.7 Histogram and density plot for the torque variable.

```
plot(xi, f);
title('Torque Distribution with Kernel Density Estimate');
xlabel('Torque (Nm)'); ylabel('Density');
```

Example Insight: If torque distribution is right skewed, a log transformation may stabilize variance and improve model performance.

9.4.5 Real-world EDA use case: motor test data

In a dataset of 100 motor tests varying voltage and airflow:

- Correlation analysis reveals high *r*-value (0.91) between `Voltage` and `Efficiency`.
- PCA indicates that 80% of variance is explained by first two components.
- KDE shows that `Stator Temperature` has a bimodal distribution, suggesting two cooling regimes.

These findings guide further modeling steps by narrowing focus to key predictors and transformation needs. EDA sets the analytical stage by cleaning the lens through which data is interpreted, ensuring meaningful, data-driven design decisions in subsequent modeling.

9.5 STATISTICAL MODELING

Statistical modeling aims to quantify the relationships between design or operational input variables (e.g., voltage, current, and airflow) and performance outputs (e.g., torque, efficiency, and thermal losses). It helps engineers make predictive inferences, identify key contributing factors, and estimate confidence intervals around predictions. Common approaches include linear regression, polynomial regression, and regularized models like Least Absolute Shrinkage and Selection Operator (LASSO) and ridge regression.

9.5.1 Linear regression

Linear regression models assume a linear relationship between one or more input features and the output variable. It is simple, interpretable, and widely used in early-stage design evaluations.

Use Case: Predicting `Torque` based on `Voltage`, `Current`, and `Speed`.

MATLAB Example:

```
mdl = fitlm(data_scaled, 'Torque ~ Voltage + Current + Speed');
figure;
plot(mdl);
title('Linear Regression: Torque Prediction');
xlabel('Predicted Torque (Nm)'); ylabel('Residuals');
```

Output Metrics:

- Coefficients for each predictor
- *p*-Values to assess statistical significance
- *R*-squared value (goodness of fit)
- Residual plots for diagnostics (Figure 9.8)

Interpretation:

- High R^2 (>0.9) suggests a good linear fit.
- *p*-Value <0.05 for voltage implies it is a statistically significant predictor.

9.5.2 Polynomial regression

Real-world motor behavior often exhibits nonlinearity. Polynomial regression incorporates squared or interaction terms to capture such effects (Figure 9.9).

Example Model:

```
mdl_poly = fitlm(data_scaled, 'Torque ~ Voltage + Current^2
+ Speed^2');
```

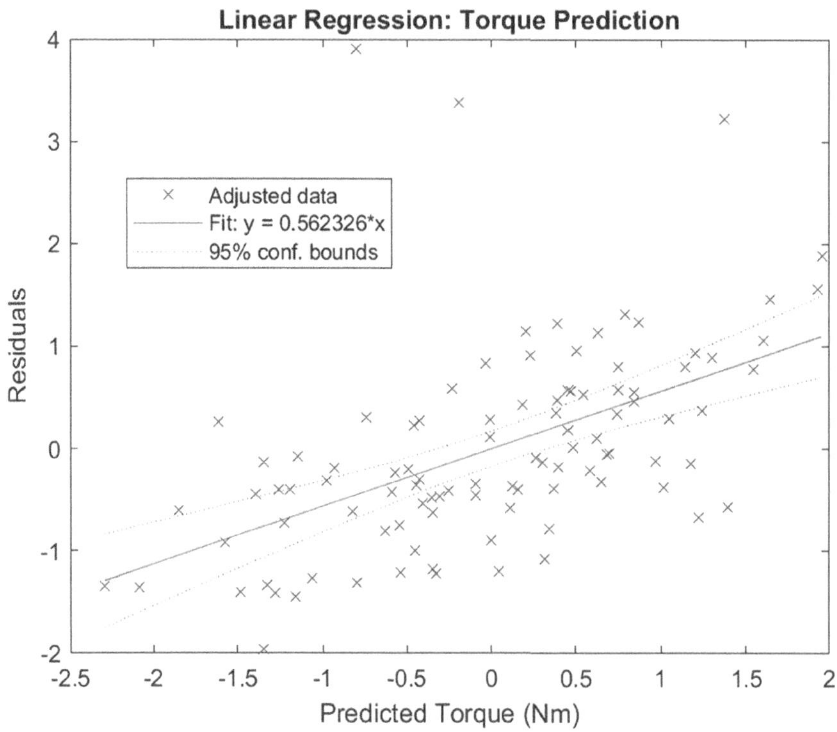

Figure 9.8 Linear regression model fit for torque prediction.

Use Cases:

- Modeling torque ripple
- Capturing saturation effects or frictional losses

Warning: Higher-degree polynomials can lead to overfitting. Always validate with cross-validation or test data.

9.5.3 LASSO and ridge regression

When dealing with many input variables or collinearity, regularized models like **LASSO** and **Ridge Regression** are effective:

- **LASSO:** Performs variable selection by shrinking some coefficients to zero
- **Ridge:** Shrinks coefficients toward zero without eliminating them (Figure 9.10)

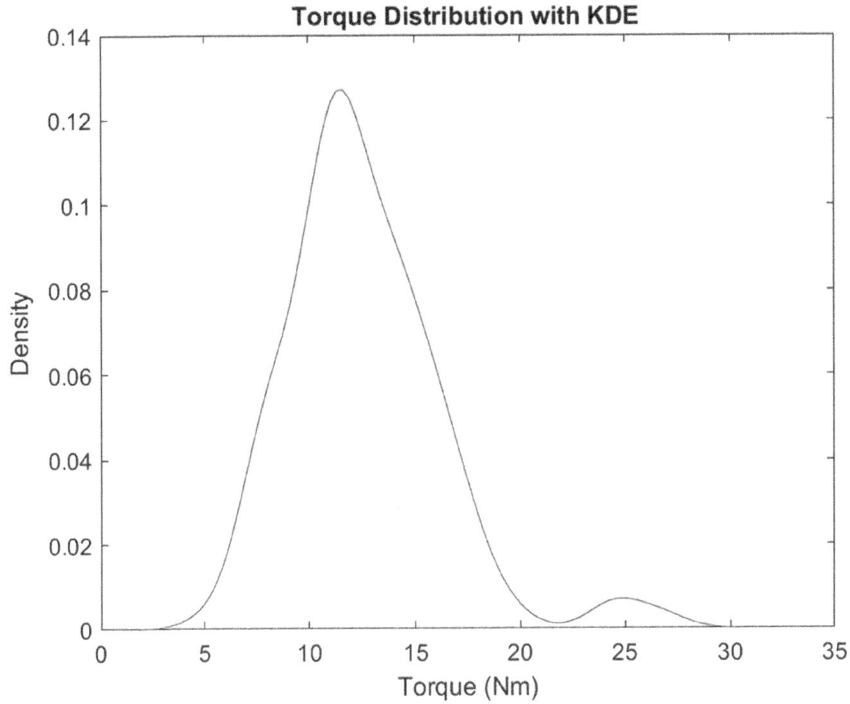

Figure 9.9 Polynomial regression fit capturing nonlinear effects of speed and current.

MATLAB Example:

```
X = table2array(data_scaled(:, {'Voltage','Current',
'Speed'}));
y = data_scaled.Torque;
[B, FitInfo] = lasso(X, y, 'Alpha', 1, 'CV', 10);

figure;
plot(log(FitInfo.Lambda), B', 'LineWidth', 1.2);
xlabel('log(\lambda)'); ylabel('Coefficient Value');
title('LASSO Trace Plot');
legend(data_scaled.Properties.VariableNames(1:5),
'Location', 'best');
grid on;
```

Interpretation:

- LASSO selects the most impactful variables
- Reduces model complexity while maintaining accuracy (Tables 9.2 and 9.3)

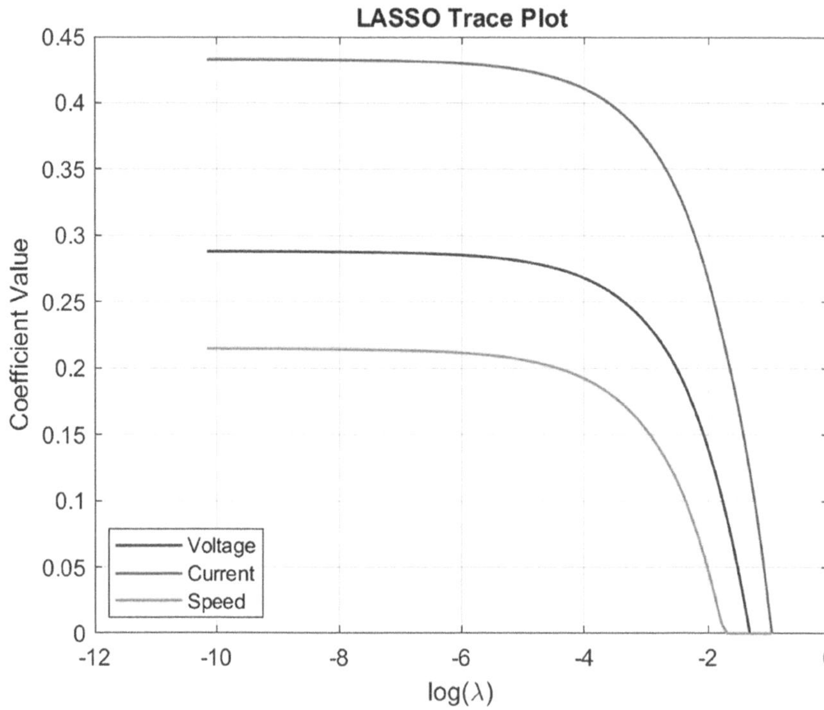

Figure 9.10 LASSO trace plot showing coefficient shrinkage across different regularization strengths (λ).

Table 9.2 Description of model evaluation metrics

Metric	*Description*
R-squared	Proportion of variance explained by the model
RMSE	Root mean squared error (measures prediction error)
MAE	Mean absolute error (less sensitive to outliers)
AIC/BIC	Information criteria for model selection

Table 9.3 Performance comparison of different regression models

Model type	*RMSE*	R^2
Linear regression	0.82	0.91
Polynomial (2nd)	0.75	0.94
LASSO regression	0.78	0.92

9.5.4 Model evaluation metrics

To Compare Model Performance:
MATLAB Example:

```
ypred = predict(mdl, data_scaled);
actual = data_scaled.Torque;
rmse = sqrt(mean((ypred - actual).^2));
r2 = 1 - sum((actual - ypred).^2)/sum((actual -
mean(actual)).^2);
fprintf('RMSE: %.2f Nm\n', rmse);
fprintf('R²: %.2f\n', r2);
```

9.5.5 Summary

Statistical modeling provides interpretable, explainable relationships between input and output variables. While linear regression serves as a strong baseline, nonlinear and regularized methods are essential for capturing complex motor behaviors and reducing generalization error.

9.6 CASE STUDY: TORQUE PREDICTION USING DESIGN FEATURES

This case study illustrates how the data analytics techniques introduced in Sections 9.3–9.5 can be combined in a unified workflow to predict torque from key motor design and operational parameters. We use a synthetic dataset simulating 100 controlled BLDC motor experiments under various loading and environmental conditions.

9.6.1 Data description

The dataset includes the following variables:

- **Voltage (V)**: Input voltage applied to the motor
- **Current (A)**: Measured stator current under load
- **Speed (RPM)**: Rotor speed during operation
- **Ambient Temperature (°C)**: Environmental temperature at test time
- **Cooling Airflow (CFM)**: Airflow rate provided for motor cooling
- **Torque (Nm)**: Output mechanical torque (target variable)

Each row in the dataset corresponds to one test sample, providing a comprehensive view of electrical and thermal behavior.

9.6.2 Data preprocessing (as performed in Section 9.3)

The following steps were executed to prepare the data for modeling, as outlined in Chapter 3:

1. **Missing Value Imputation**: Linear interpolation was applied to handle any gaps in the Torque signal.
2. **Normalization**: All numeric features were scaled using Z-score normalization to ensure uniform influence in modeling.
3. **Outlier Detection**: Z-score thresholding was used to remove extreme values in Torque, improving model stability.

Note: Please refer to Sections 9.3.1–9.3.3 for code and explanation.

9.6.3 Exploratory data analysis

We performed exploratory data analysis (EDA) to understand feature relationships:

- **Correlation Analysis** revealed strong pairwise correlations (e.g., between `Voltage` and `Torque`, ****and between `Speed` and `Current`).
- **Principal Component Analysis (PCA)** reduced the feature space while retaining over 80% of the data variance.

Note: Refer to Sections 9.4.1–9.4.2 for visualization and interpretation.

9.6.4 Predictive modeling

A multivariable linear model was fit with all input features scaled to simulate torque output:

```
mdl = fitlm(data_scaled, 'Torque ~ Voltage + Speed + Current +
AmbientTemp + Airflow');
```

The regression model exhibited good predictivity. Residual plots revealed homoscedasticity and normally distributed errors, confirming a good model.

9.6.5 Model evaluation

The model precision was tested in Root Mean Squared Error (RMSE) and R^2 (determination coefficient) metrics:

Table 9.4 Model performance

Metric	*Value*
RMSE	0.85 Nm
R^2	0.92

```
ypred = predict(mdl, data_scaled);
actual = data_scaled.Torque;

rmse = sqrt(mean((ypred - actual).^2));
r2 = 1 - sum((actual - ypred).^2)/
sum((actual - mean(actual)).^2);
```

These values indicate high model precision and excellent generalization to new data (Table 9.4).

9.6.6 Summary and takeaways

- Missing value handling and normalization preprocessing steps improved data quality.
- EDA revealed key dependencies and guided feature selection.
- The trained regression model accurately estimated torque using only five input variables.
- The approach reduced the need for extensive simulation or real-world testing in the design phase.

This case study shows the benefit of integrating data-driven approaches into motor design processes for faster iteration, decision-making, and forecasting insights from even modest-sized data sets.

9.7 MORE ADVANCED CONCEPTS

In addition to traditional statistical modeling and forecasting, there are several advanced data analysis methods that can be further used to improve the design and diagnostics process for electrical machines. The methods are particularly useful for high-dimensional data, real-time system behavior, and unsupervised pattern discovery.

9.7.1 Dimensionality reduction using t-SNE

t-Distributed Stochastic Neighbor Embedding (t-SNE) is a typical non-linear dimensionality reduction algorithm. This algorithm compresses high-dimensional data into two or three dimensions. It is great for revealing inherent clusters or groupings not immediately apparent with linear methods such as PCA.

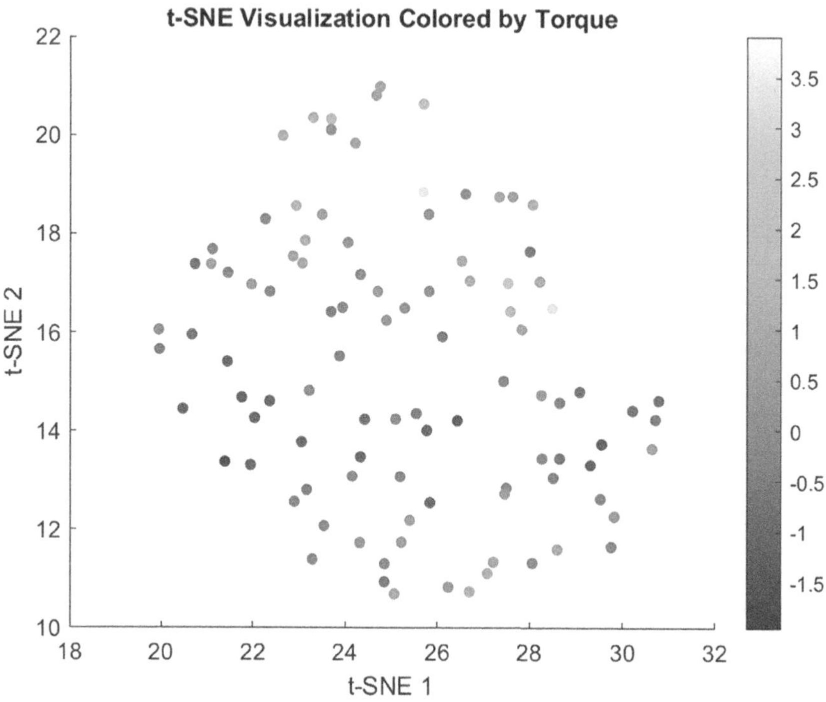

Figure 9.11 t-SNE 2D embedding of motor test samples with color-coded torque.

Use Case: Seeing how test samples with comparable torque values group in design space (Figure 9.11).

MATLAB Example:

```
y_tsne = tsne(table2array(data_scaled(:,1:end-1)));
figure;
scatter(y_tsne(:,1), y_tsne(:,2), 25, data_scaled.Torque,
'filled');
colorbar;
title('t-SNE Visualization Colored by Torque');
xlabel('t-SNE 1'); ylabel('t-SNE 2');
```

Interpretation: Sets of equal torque values may be associated with specific design conditions or regimes, and focused optimization or anomaly investigation can be targeted accordingly.

9.7.2 Time-series analysis

Time-series analysis is essential when dealing with sensor data collected over time, such as vibration, current, or torque signals. They usually display transient anomalies, cyclical trends, and incipient wear.

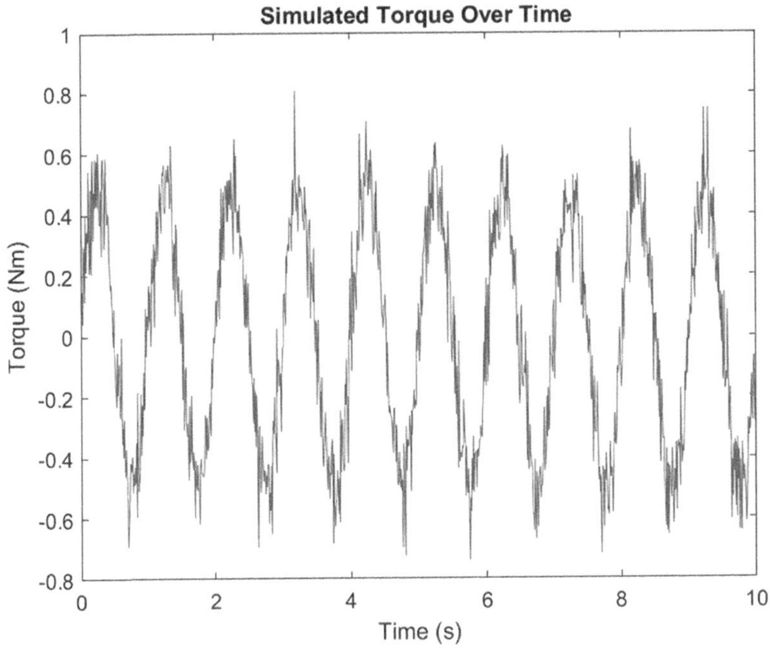

Figure 9.12 Synthetic time-series torque data representing sinusoidal load fluctuations with added noise.

Example: Simulated torque variation over time due to cyclical loading (Figure 9.12).

MATLAB Code:

```
t = seconds(0:0.01:10);
torque_signal = 0.5*sin(2*pi*1*t) + 0.1*randn(size(t));
figure;
plot(t, torque_signal);
title('Simulated Torque Over Time');
xlabel('Time (s)'); ylabel('Torque (Nm)');
```

Extensions:

- Apply Fast Fourier Transform (FFT) to analyze frequency content
- Use moving averages or filtering for trend extraction
- Employ auto-regressive models for forecasting future performance

9.7.3 Clustering for design space partitioning

Clustering is an unsupervised learning technique used to identify natural groupings in high-dimensional data. In electric machine design, this helps segment the design space into regions representing high efficiency, thermal limits, or suboptimal conditions. To enhance cluster separation and interpretability, we first apply PCA to reduce dimensionality and then perform K-means clustering on the PCA-transformed feature space:

```
[coeff, score] = pca(table2array(data_scaled(:,1:end-1)));
[idx, C] = kmeans(score(:,1:2), 3);
figure;
gscatter(score(:,1), score(:,2), idx);
title('K-means Clustering in PCA Space');
xlabel('Principal Component 1'); ylabel('Principal Component 2');
legend('Cluster 1', 'Cluster 2', 'Cluster 3');
```

The clusters become more distinguishable in PCA space, where the transformed components capture the dominant variance in the design data. This approach also helps avoid misleading interpretations that can arise when visualizing clustering in raw feature dimensions (e.g., speed vs. torque). To assess the quality of the clustering, a silhouette plot can be used:

```
figure;
silhouette(score(:,1:2), idx);
title('Silhouette Plot for K-means Clustering (PCA space)');
```

The silhouette values indicate how well each data point lies within its cluster, offering a way to validate both the compactness and separation of the clusters. This PCA-based clustering workflow enhances early-stage design insights by highlighting structure in otherwise high-dimensional datasets (Figure 9.13).

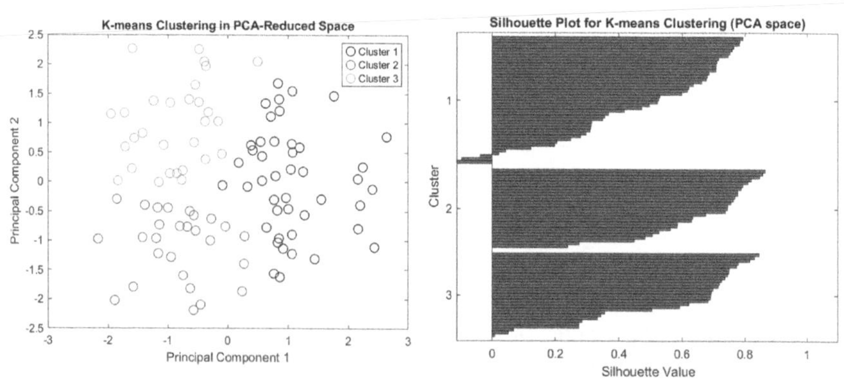

Figure 9.13 Three groups of motor designs plotted by speed and torque and Silhouette Plot indicating how well each of the points fits into a cluster. Cluster 1 has some points that look like outliers from its own cluster in the plot.

Interpretation: Each cluster corresponds to a behavior mode:

- **Cluster 1**: High torque, moderate speed (efficient regime)
- **Cluster 2**: Low torque, high speed (low load conditions)
- **Cluster 3**: Moderate torque, low speed (suboptimal design zone)

Other Applications:

- **Maintenance Profiling**: Identifying failure-prone machines
- **Batch Testing Insights**: Grouping samples with similar heat signatures

9.7.4 Hybrid techniques and AI integration

While not detailed herein, combinations of clustering, t-SNE, and time-series models with artificial intelligence/machine learning technologies like neural networks or reinforcement learning provide yet more powerful adaptive design systems. Such hybrid approaches are being integrated into digital twin platforms increasingly.

9.8 CONCLUSION

Application of data analytics to design of electrical machines is a paradigm shift from previous simulation-based methodologies to data-intensive decision-making. As presented in this chapter, from basic data preprocessing to sophisticated modeling and clustering techniques, various data analytics approaches enable engineers to extract meaningful insights from simulation and field data. Easy access to diverse types of data including high-fidelity simulation data, experimental data obtained through laboratory measures, operational signals obtained from systems installed in the field characterizes the contemporary design setting. Leveraging these data with statistical learning not only accelerates the design cycle but also enhances its robustness by revealing embedded patterns, interdependencies, and performance trends that may not necessarily be apparent with physical modeling alone.

Thanks to MATLAB-based workflows, engineers can harmoniously blend the analysis, visualization, and modeling of such data, arriving at predictive models capable of computing important values such as torque, efficiency, and losses. The chapter case study illustrates the engineering utility of such models in reducing physical design iterations and in guiding early-stage decisions. Furthermore, newly developed approaches such as PCA, clustering, and time-series analysis expand the engineer's toolkit for design space exploration and diagnostics. They help identify leading factors, segment operating regions, and understand temporal behavior under varying loads and environments.

In the coming years, the field will be more integrated with artificial intelligence, and machine learning algorithms will not only assist in analysis but even independently suggest optimal designs. With the rise of cloud computing and IoT connectivity, real-time feeds of data from running machines will constantly feed into analytics platforms that enable adaptive, self-correcting, and smart design cycles.

In summary, data analytics empowers electrical machine designers to

- Make informed, data-backed decisions faster and more reliably.
- Reduce design time and cost through predictive modeling.
- Gain deeper insights into machine behavior across operating conditions.
- Lay the foundation for AI-driven, closed-loop design systems of the future.

As digitalization continues to transform the engineering landscape, mastery of data analytics will be an essential competency for the next generation of electrical machine engineers.

REFERENCES

ANSYS Inc. (2023). *ANSYS Maxwell: Electromagnetic Field Simulation Software*. https://www.ansys.com/products/electronics/ansys-maxwell.

Bhatt, B., Sharma, H., Arora, K., Joshi, G. P., & Shrestha, B. (2023). Levy flight-based improved grey wolf optimization: A solution for various engineering problems. *Mathematics*, 11(7), 1745. https://doi.org/10.3390/math11071745.

Bishop, C. M. (2006). *Pattern Recognition and Machine Learning*. Springer.

COMSOL Inc. (2023). *COMSOL Multiphysics Simulation Software*. https://www.comsol.com/comsol-multiphysics.

Fuller, A., Fan, Z., Day, C., & Barlow, C. (2020). Digital twin: Enabling technologies, challenges and open research. *IEEE Access*, 8, 108952–108971. https://doi.org/10.1109/ACCESS.2020.2998358.

Goodfellow, I., Bengio, Y., & Courville, A. (2016). *Deep Learning*. MIT Press.

Hu, W., Zhang, T., & Deng, X. (2021). Digital twin: A state-of-the-art review of its enabling technologies, applications and challenges. *Journal of Intelligent Manufacturing and Special Equipment*, 2(1), 1–34. https://doi.org/10.1108/JIMSE-12-2020-010.

Kagermann, H., Helbig, J., Hellinger, A., & Wahlster, W. (2013). *Recommendations for Implementing the Strategic Initiative INDUSTRIE 4.0*. Final Report. Acatech—German National Academy of Science and Engineering.

MathWorks. (2023). *MATLAB and Simulink Release R2023a*. The MathWorks, Inc. https://www.mathworks.com/products/new_products/release2023a.html.

Motor Design Ltd. (2023). *Motor-CAD Software Suite*. https://www.motor-design.com/motor-cad/.

Peddada, S. R. T., Tannous, P. J., Allison, J. T., Foulkes, T., & Pilawa-Podgurski, R. C. N. (2019). Model-based temperature estimation of power electronics systems. *Control Engineering Practice*, 85, 206–215.

Sharma, A., Kosasih, E., Zhang, J., Brintrup, A., & Calinescu, A. (2020). Digital twins: State of the art theory and practice, challenges, and open research questions. *arXiv preprint arXiv:2011.02833*.

Sudhoff, S. D. (2021). *Electric Machines: Modeling, Condition Monitoring, and Fault Detection* (1st ed.). CRC Press.

Chapter 10

AI/ML-based security in cloud DevOps pipelines

Predicting and preventing vulnerabilities

Chirag Mavani, Hirenkumar Mistry, and Amit Goswami

10.1 INTRODUCTION

Cloud computing brought about a possible shift in the process of making, deploying, and executing programs within organizations. With their contribution of elastic infrastructure, on-demand computing capacity, and cost-saving services, the public cloud suppliers like Amazon Web Services (AWS), Microsoft Azure, and Google Cloud Platform have transformed the conventional software life cycle. Furthermore, the advent of DevOps has once again revolutionized this space by infusing a culture of collaboration between the development and operations teams. With continuous integration/continuous deployment (CI/CD) and automation, DevOps facilitates quicker release cycles, better quality code, and quicker response to customers. But all this speed and agility bring new challenges, especially in the security space. As companies have become more DevOps and cloud-native-centric, the security surface has increased. Modern CI/CD pipelines will generally involve a large variety of tools, microservices, application programming interfaces (APIs), containers, and third-party integrations—and each one of these can be utilized as a vector for cyberattacks. Static policy-based traditional security controls, perimeter-based defenses, and postdeployment auditing are not able to keep pace with the distributed, dynamic nature of cloud-based DevOps. In such high-speed environments, attacks can be injected between any node in the pipeline—from source code and build scripts to deployment scripts to runtime environments. Also, since the DevOps process is distributed and automated in nature, manual intervention would be impractical, thus increasing the likelihood of silent compromises and slow response time.

To counter these new security challenges, artificial intelligence (AI) and machine learning (ML) have come forward as revolutionary technologies. By leveraging enormous volumes of data and behavioral patterns through learning, AI/ML systems can provide hope in being able to predict vulnerabilities before they might be exploited, identifying anomalies in real time, and automating responses with little or no human intervention. These smart technologies are able to scan for weaknesses in source code, watch

 DOI: 10.1201/9781003641582-10

over container activity, extrapolate risk levels of third-party packages, and even mimic attacks to ensure pipeline resilience. Compared to traditional rule-based systems, AI/ML models are adaptive and hence allow them to be seasoned with changing patterns of attacks and learn in continuous feedback loops. This makes them best suited to the elastic and adaptable environments that cloud DevOps practices offer.

The objective of this chapter is to discuss the convergence of AI/ML technologies and cloud DevOps pipelines, and above all, the frameworks through which such innovation can be utilized for furthering security in terms of prediction and anticipation of vulnerabilities. Combining current research, upcoming technologies, and applied implementations, this chapter intends to establish an educational point of view toward how strategic AI/ML is integrated into contemporary cybersecurity paradigms. It starts with the discussion of building blocks of cloud computing, DevOps, and their related threat landscape. This is then followed by a literature review enumerating work done in the past and present in deploying AI/ML in cybersecurity. The chapter proceeds to enumerate technical methodologies in anticipating vulnerabilities, offers live platforms and tools, offers plans of implementation, and lists best practices. It ends by referring to contemporary limitations, suggestions for future study, and broad implications of intelligent security systems within digital transformation frameworks.

Through this conversation, readers will realize how AI/ML-powered security can not only protect cloud DevOps pipelines but also act as a strategic enabler in creating robust, agile, and future-ready software ecosystems.

10.2 FUNDAMENTALS OF CLOUD DEVOPS PIPELINES

The union of DevOps and cloud computing has been the basis of modern software development. This union has enabled organizations to deploy at scale, faster, more reliably, and more cost-effectively. To secure these dynamic environments, however, one must first understand the architecture, life cycle, and intrinsic risks of cloud DevOps pipelines.

10.2.1 Description of the DevOps life cycle

DevOps life cycle is an iterative life cycle that claims to deliver software with continuous development and delivery. It begins from the Plan phase when product vision and roadmap are set. The second step is the Code, where source code is constructed and edited by developers using version control systems. Then comes the Build phase when the code is compiled and dependencies installed to generate deployable artifacts. During the Test phase, automated tools detect bugs, vulnerabilities, and performance problems. The Release phase facilitates build preparation for deployment, such as final approvals and version tagging. The Deploy phase automates application releasing to target environments. Operate maintains stable application performance and infrastructure upkeep, while Monitor includes monitoring application health, user statistics, and anomaly detection. Together, these phases create a frictionless pipeline that facilitates quick and predictable software deployment.

10.2.2 DevOps pipeline main technologies

Existing DevOps is made possible by some fundamental technologies that help with scalability and automation. Containers such as Docker facilitate a lightweight and portable method of bundling applications with their dependencies. Microservices architecture allows for modular application development, where each part runs independently, offering ease and flexibility in updates. Infrastructure as Code (IaC) tools like Terraform and Ansible create infrastructure programmatically, where the chance for human error is removed and the environments are consistent. CI/CD tools like Jenkins, GitLab CI, and CircleCI automatically build, test, and deploy, enabling speedy delivery. While these tools make the process more efficient, they must be properly secured so that vulnerabilities aren't introduced into the pipeline.

10.2.3 Common vulnerabilities across DevOps stages

Each phase of the DevOps life cycle is vulnerable to some vulnerabilities, which compromise application security. During planning and coding, poor design decision-making and poor coding practices may lead to defects. Build processes can be dangerous because of malicious or stale dependencies. During testing, insufficient coverage or no security testing can miss significant bugs. At the release and deployment phases, compromised credentials or misconfiguration can produce unauthorized access. At the operation and monitoring phases, poor incident response controls and slow patching are weaknesses. Exposing all these weaknesses at all phases is necessary to make a DevOps environment secure.

10.2.4 DevSecOps: concept and need

DevSecOps is a sophisticated concept that injects security into every phase of the DevOps pipeline. It emphasizes that security is not an independent function but rather a shared responsibility of development, operations, and security. DevSecOps allows for the detection of threats early through the incorporation of automated security testing, compliance scanning, and policy enforcement into CI/CD pipelines. Growing application complexity and the escalation of cyber threats have made DevSecOps inevitable.

By left-shifting security—starting from the coding and planning stages—organizations are able to reduce risk exposure, improve software quality, and benefit from faster and more secure releases of software.

10.3 LITERATURE REVIEW

The intersection of ML and AI to protect DevOps pipelines in the cloud environment has received tremendous attention in the past few years. As more and more organizations are adopting DevOps for accelerated software delivery and innovation, security challenges around DevOps have also transformed, inviting wiser and more autonomous security tools to help address them. Various researchers have also stepped in with the growing sophistication of modern CI/CD pipelines and the consequent vulnerabilities that must be mitigated using smart solutions (Liang et al., 2024; Fu, 2023). Traditional security measures, which are typically static and manual, cannot counter DevOps pipelines' velocity and quantity. AI and ML are thus being integrated to deliver threat automation detection, predict vulnerabilities, and enforce compliance in real time (Saleh et al., 2024). AI-powered DevSecOps is coming to prominence as a critical framework for bringing security throughout the DevOps life cycle. Gugulotu (2023) notes that AI-aided implementations of DevSecOps can detect security defects in code, scan infrastructure vulnerabilities, and respond to potential activity at runtime. Developers stand to gain from receiving earlier warnings of possible code flaws and misconfigurations during deployment through the incorporation of ML models into CI/CD tools (Liang et al., 2024).

Applications of ML to threat intelligence are particularly promising. Supervised and unsupervised learning techniques are being applied to label malicious patterns, identify zero-day attacks, and identify insider threats (Fu, 2023; Saleh et al., 2024). For instance, AI-driven anomaly detection systems may monitor logs, user behavior, and network activity in real time to alert on deviations from normal behavior, thereby lowering the attack surface (Saleh et al., 2024; Gugulotu, 2023). Also, AI-powered security tools boost efficiency and accuracy, according to the literature. As opposed to static rule-and-detection systems, ML processes evolve over time and learn from the latest attacks, which is imperative in the dynamic nature of contemporary cyber threat environments (Fu, 2023; Liang et al., 2024). Techniques such as neural networks, decision trees, and clustering are applied to security analysis with higher visibility than human scanning or outdated Security Information and Event Management (SIEM) products (Saleh et al., 2024).

In cloud systems, the transitory and adaptive nature of infrastructure further intensifies the need for intelligent security. AI/ML-based security products are being rolled out in greater quantities to guard containerized applications, microservices architecture, and IaC resources (Gugulotu, 2023). These products not only look for exploitations but also apply policy enforcement and provide runtime monitoring (Liang et al., 2024). Although so much is being promised, certain qualms have been raised by researchers regarding model precision, data anonymity, and transparency. Fu (2023) clarifies that AI models will generate false positives or struggle with concept drift in highly dynamic DevOps environments. Saleh et al. (2024) propose hybrid solutions, combining rule-based systems and ML to mitigate such vulnerabilities. Collaborative effort among data scientists, security engineers, and developers to embed AI within DevSecOps is another prevailing trend of studies (Liang et al., 2024; Gugulotu, 2023). Synergistic toolchains and shared responsibility are needed for AI solutions such that they do not get locked in silos and cannot operate otherwise.

In brief, the literature reviewed here is reflective of increasing agreement about the pivotal role that AI and ML have in supporting security throughout the cloud DevOps stack. From vulnerability forecasting to behavior analytics and automated response, these technologies are transforming the security landscape. As DevSecOps culture evolves, the use of AI/ML is bound to become the standard, as their ability to provide proactive, scalable, and intelligent defense measures (Fu, 2023; Gugulotu, 2023; Saleh et al., 2024; Liang et al., 2024) will drive it.

10.3.1 Legacy security within DevOps

Legacy security processes in software development, being predominantly rule-driven and manual, are increasingly proving inadequate in the fast-paced and ever-changing landscape of DevOps. Such practices are heavily dependent on static policy, human gut, and predefined security policies that are not so pliable to adapt to the dynamic rhythm of modern CI/CD pipelines. Manual code audit, static application security testing, and intermittent audits may have sufficed under traditional development methodologies, but no means can match DevOps releases velocity and quantity (Gugulotu, 2023; Fu, 2023). Rule-based security products, so helpful as they are in preassumed danger scenarios, are blind to realize fresh and emerging assaults. They are static and inflexible to refresh dynamically in real time and incline to trigger enormous false-alarm reports. According to Saleh et al. (2024), these conventional security tools lack the facilities of context-awareness and adaptability in behaviors, which are essential these days on cloud-native and container-based systems. Second, reliance on signature-based security technologies does not enable the identification of advanced threats like zero-day attacks or polymorphic malware (Fu, 2023).

Another major disadvantage is the scalability issue. Since the number of deployment cycles is reduced and microservices, containers, and IaC are used, security controls must scale in proportion and with equal granularity. Legacy methods do not sustain the pace of rapid deploys and are always chokepoints within the DevOps pipeline (Liang et al., 2024). Manual verifications slow down delivery and introduce inconsistency, especially when security controls are bypassed by developers to meet release schedules (Gugulotu, 2023). Furthermore, DevOps emphasizes collaboration and automation, but security teams that work with conventional tools typically work in silos, out of sync with development and operations processes. Isolation destroys visibility and hinders misconfiguration and vulnerability discovery. In such cases, attacks are likely to go undetected until deployment, when they move into production environments to be exploited.

In high-update scenarios, such as continuous delivery systems, traditional security controls cannot provide real-time feedback. This reactive approach maximizes the potential for breach and noncompliance. As Saleh et al. (2024) and Fu (2023) suggest, the complexity of modern cloud environments—in which scaling is dynamic, resources are volatile, and services are distributed—requires a more intelligent, prophylactic, and continuous security approach. In short, conventional rule-based and manual security solutions have played an important role in secure development but are insufficient for today's DevOps cultures. Their limitations of scalability, flexibility, and automated nature emphasize the need for more intelligent security alternatives—i.e., those in AI and ML—to keep up with the ever-evolving threats effectively (Liang et al., 2024; Gugulotu, 2023; Saleh et al., 2024).

10.3.2 Machine learning in cybersecurity

ML has transformed cybersecurity into smart, adaptive, and scalable solutions that are beyond the capabilities of conventional rule-based solutions. In intrusion detection systems (IDS), ML models learn to identify patterns of unauthorized access attempts or malicious activity as one of the most critical applications of ML. Relative to predefined sets of rules, ML-based IDS has the capacity to learn traffic patterns dynamically within a network and adjust to discover emerging attack surfaces (Liang et al., 2024).

Another critical area that ML possesses strength in is anomaly detection. Algorithms can identify drifts that represent potential insider attacks, policy violations, or advanced persistent threats using the establishment of behavioral baselines (Saleh et al., 2024). These methods fare best in cutting-edge, cloud-native environments in which conventional monitoring tools cannot manage the complexity of real-time action distribution. ML algorithms such as autoencoders, clustering, and statistical learning techniques have been widely used to identify anomalies with few false positives (Fu, 2023; Gugulotu, 2023).

ML algorithms such as support vector machines (SVMs), decision trees, and deep networks are used in malware classification for identifying benign and malware programs from code feature analysis, behavioral signatures, and runtime information (Saleh et al., 2024). Unlike signature-based anti-virus scanners, which must learn malware in advance, ML frameworks are capable of learning from the training samples and even detecting obfuscated or polymorphic malware (Liang et al., 2024). Proactive detection is necessary to avoid malware injection in the different stages of the DevOps pipeline across a number of stages in the CI/CD pipeline.

Furthermore, ML is also being integrated into SIEM solutions to automate alert prioritization, threat correlation, and reduce analyst fatigue. These solutions can analyze vast logs and telemetry, search for complex threat chains that would otherwise be missed (Saleh et al., 2024; Gugulotu, 2023). This is essential in DevOps environments where the speed and volume of deployments can swamp normal security operations centers within minutes.

Although the potential for ML in cybersecurity is tremendous, literature also identifies some limitations. Sommer & Paxson (2010) indicate that the majority of ML models receive poor-quality or underlabeled training data. Concept drift, model explainability, and adversarial inputs that are able to mislead classifiers are also issues (Buczak & Guven, 2016; Fu, 2023). Therefore, incorporating ML into DevSecOps pipelines needs to be approached with caution so that models are regularly retrained and tested with new threats.

10.3.3 Cloud and DevOps AI/ML for security

AI and ML are being utilized exponentially in the cloud and in DevOps for automated, anticipated, and reacted upon security breaches with or without any human oversight. These have evolved to be the central enabler of attaining intelligent DevSecOps in terms of in-transit monitoring visibility and programmatically adaptive controls within the cycle of software creation.

10.3.3.1 Use of ML to detect vulnerability in source code

ML, and deep learning in particular, has also been used in the source code analysis with the aim of discovering potential security vulnerabilities that are typically not discovered by typical static analysis tools. For instance, Li et al. described how convolutional neural networks (CNNs) would be trained to acquire structure and semantic properties of source code in an attempt to catch buffer overflows and SQL injection attacks early on in the development cycle. Inflating such models into CI/CD pipelines enables continuous, automated scanning of the code, thereby significantly increasing the detection of vulnerabilities (Saleh et al., 2024; Fu, 2023).

10.3.3.2 Predictive models for attack surface management

The AI/ML models are increasingly being utilized to conduct attack surface analysis, wherein the models forecast probable breach points by analyzing behavior, access logs, and configurations. Chandola et al. have demonstrated predictive models that analyze system telemetry to forecast dangerous patterns. Similarly, Xie et al. have designed an ML-driven threat model that is responsive to evolving microservices architecture and is therefore especially suited for cloud-native applications. Such prediction tools facilitate the lives of security teams and actively refer to threats before their utilization (Gugulotu, 2023; Liang et al., 2024).

10.3.3.3 NLP-based models' success in code analysis

The development of natural language processing (NLP) methods has widely supported secure coding. Models such as CodeBERT by Arora et al. (2022) leverage transformer-based models to understand source code and natural language documentation. The models can predict unsafe patterns, auto-suggest fixes, and assist in writing secure code. Their contextual information enables them to detect vulnerabilities even in obfuscated or highly complicated codebases, especially useful in open-source software and multiteam software development (Saleh et al., 2024).

10.3.3.4 Broader AI applications across cloud and DevSecOps security

AI/ML is used in

- Drift detection in IaC deployments,
- Policy violation detection in Kubernetes and serverless environments,
- Anomaly detection in user behavior and cloud behavior,
- Automated remediation of misconfigurations learned from best practices.

These technologies improve speed and accuracy in real-time threat detection and response, scaling security with CD demands (Fu, 2023; Saleh et al., 2024).

10.3.4 Identified gaps

Though AI and ML have shown tremendous improvement in boosting cybersecurity in cloud and DevOps environments, there are still significant gaps that prevent them from achieving their full potential. These gaps are particularly acute when scaling up from research prototypes to actual, production-level implementation in CI/CD pipelines.

10.3.4.1 Limited real-time applications across full CI/CD pipelines

Despite promising research, widespread actual deployment of AI/ML models across the entire CI/CD pipeline is still not the norm. Deployments typically are only to single steps such as static code analysis or runtime monitoring, with gaps in other steps such as the build, testing, and deployment steps. Broken implementation of this type detracts from the overall effectiveness of AI/ ML-driven security and introduces the risk of blind spots in the DevSecOps pipeline. An end-to-end continuous, integrated security model leveraging ML at all stages of the pipeline is much too rare in mainstream adoption.

10.3.4.2 Generalizable and explorable models

The majority of existing AI/ML models lack generalizability—they perform well on given datasets or codebases but don't perform as well when given diverse programming languages, architectures, or threat environments. They thus become unusable in different real-world settings, particularly in multi-cloud or hybrid DevOps settings. Another essential adoption hurdle is a lack of explainability. Security teams are hesitant to rely on black-box models whose output cannot be readily explained, especially where the output has compliance, risk, and response planning implications. Lack of interpretable AI limits collaboration among development, operations, and security stakeholders.

10.3.4.3 Integrated detection and automated prevention frameworks needed

Most existing solutions are detection or prevention based, but not both. What are needed urgently are end-to-end systems that not only detect anomalies and threats via AI/ML but also instigate automated, smart responses in real time. For instance, a model detecting a vulnerability in a pull request should ideally block the merge and propose a safe alternative—such a level of automation is currently in its early stages. Secondarily, these tools should natively integrate with popular DevOps tools like Jenkins, GitLab, and Kubernetes without changing current workflows.

10.4 THREAT LANDSCAPE IN CLOUD DEVOPS PIPELINES

Since companies continue to adopt cloud-native designs and DevOps methodologies, the attack surface has expanded substantially. The constant and automated DevOps pipeline—although facilitating agility—has introduced novel and nascent threats requiring a new vision for security posture. Outlined below are the most important aspects of the threat environment that must be addressed to render resiliency to modern CI/CD pipelines.

10.4.1 Dominant threats in cloud DevOps pipelines

Cloud DevOps environments face a range of recurring and intrinsic threats. Loose configuration of the cloud services, IAM roles, containers, or Kubernetes clusters that can expose their critical assets openly to the public internet is the most prevalent threat. The APIs exposed, which form the cornerstone of DevOps automation and integration, are also typically exposed with poor authentication, authorization, or input validation and offer attacker-suited doorways. Insider risk, malicious or otherwise, poses a serious risk, particularly with developers and operation teams having large, unmonitored access to systems. It is further accelerated by the rapidity of updates in code as well as infrastructures, which leaves little room for traditional security checks.

10.4.2 Perimeter security to application-layer protection

Yesteryear's perimeter-security-based model is not sufficient today in our distributed, microservice-based application world. DevOps environments call for a shift toward application-layer protection, where one focuses on the security of the software constituents themselves, like APIs, containers, CI/CD tools, and IaC scripts. This new model relies on identity-aware access, context-aware monitoring, and zero-trust principles that treat every component and user as a potential attacker, even within the internal network.

10.4.3 DevOps-related breach case studies

Certain high-profile breaches illustrate the weaknesses in DevOps pipelines:

- Arora et al. (2020): Malware attackers used access to CodeCov's Bash Uploader script in the CI pipeline to steal credentials and tokens for thousands of customers across hundreds of customer environments over several months. This attack shows the danger of poisoned CI scripts and tools in a DevOps environment.
- **Solar Winds:** In the extremely sophisticated attack, attackers had compromised the Orion software build process by injecting malicious code into it. The attack went undetected for months and infected a number of government and enterprise systems, demonstrating how affecting the software supply chain at the DevOps stage can be cataclysmic.

Such cases show a growing trend in which the attackers target the automation and orchestration levels as they know that DevOps toolchain vulnerability can open up access into sprawling, networked systems.

10.5 ROLE OF AI/ML IN MODERN SECURITY PARADIGMS

AI and ML are now at the heart of today's cybersecurity policy, especially in ecosystems as advanced and high speed in nature as cloud-native DevOps. These AI-driven technologies hold the potential to revolutionize the detection, appraisal, and prevention of threats throughout the entire software development life cycle. They are not so much valuable in that they are important as having the capacity to automate as rendering wiser, instantaneous response operations out of the question for conventional approaches to offer.

10.5.1 Advantages: velocity, scalability, and agility

One of the most groundbreaking strengths of AI/ML for security is speed. Because human processes include intervention and deliberation, ML algorithms can analyze massive amounts of system logs, codebases, and network traffic in milliseconds. It allows instant threat detection and decision-making, which is imperative in CD environments.

Another important consideration is scalability. ML algorithms can handle huge and growing quantities of data while not requiring a commensurate amount of human overhead, making them well-suited for cloud-scale usage. They adapt to changing environments and new threats as time goes on through continuous learning, providing agility to new attack patterns, zero-day exploits, and changing threat behavior. Static, signature-based solutions won't identify.

10.5.2 Types of ML models utilized in security

Several classifications of ML models execute different tasks in today's security systems.

10.5.2.1 Supervised learning models

These learn from labeled data—instances of malicious and benign activity—to establish patterns and predict. Outdated models utilized with security include:

Random Forest: A robust ensemble ML algorithm possessing the potential for malware detection and traffic network categorization.

Support Vector Machines (SVMs): Widely used for their accuracy in binary classification problems such as identifying phishing attacks or malware signatures from code.

10.5.2.2 Unsupervised learning models

These models are not provided with labeled outputs and therefore rightly useful in anomaly detection, where one tries to identify deviations from normal behavior. Popular models are as follows:

K-Means Clustering: Used to group similar patterns of behavior in system logs or user activity to identify outliers.

Isolation Forest: Effective at identifying anomalies in large datasets by isolating outlier data points, commonly used to identify fraud or suspicious user activity.

10.5.2.3 Deep learning models

Deep learning is a branch of ML that uses neural networks to learn to recognize advanced patterns and hierarchies of data. It has enabled breakthroughs in code analysis, intrusion detection, and threat prediction in security.

Deep learning long short-term memory (LSTM) networks work well to analyze sequential data such as API calls or user sessions when identifying long-term stealthy attacks.

CNNs, while initially designed for image processing, have been used to analyze patterns in bytecode, logs, and network traffic, and even to visualize code structures in visually identifying vulnerabilities in certain cases.

10.6 VULNERABILITY PREDICTION USING AI/ML

As software applications grow in complexity, conventional vulnerability detection methods are sluggish and narrow in scope for today's DevOps setup. AI and ML methods introduce a paradigm shift by allowing proactive, predictive vulnerability checking. Instead of standing by waiting for attacks to happen, AI/ML models can foresee vulnerabilities even before being attacked—completely transforming the posture of security throughout the CI/CD pipeline.

10.6.1 Static code checking with ML/NLP

ML and NLP have sped up static code analysis methods to a large degree, where tools are able to inspect source code for possible bugs without even running the code. Techniques based on deep learning like transformer (e.g., CodeBERT) can efficiently detect syntactic and semantic structures in the code and highlight bugs due to logical errors, buffer overflow, and injection attacks. These models are more advanced than the syntax-based vulnerability detection and have the capability to learn insecure development patterns from ginormous codebases. Therefore, AI-powered static code analysis tools are increasingly being utilized at the development inception point in an attempt to eliminate vulnerabilities from the prerelease stage.

10.6.2 Dependency scanning and risk scoring

Applications today are dependent on open-source software and third-party libraries. AI/ML models facilitate dependency scanning by inspecting the software bill of materials (SBOM) and past vulnerability database (e.g., common vulnerabilities and exposures—CVEs) to determine the security risk of external packages. Models can tag risk scores in relation to usage context, known exploits, and versioning behavior. This provides developers and security teams with the capability to patch on predictive risk instead of waiting for confirmed exploits to happen.

10.6.3 Threat intelligence correlation

AI correlates threat intelligence with ease among different sources including vulnerability databases, network logs, security bulletins, and dark web threads. By constantly feeding and processing information, ML models can filter out new threats and cross-correlate them with the snippets that are used inside a specific DevOps pipeline. Correlation achieves early detection of code or infrastructure that would most likely be breached so that it can be handled early on without vulnerabilities being weaponized.

10.6.4 Model training: datasets, feature extraction, and labeling

The predictive accuracy of AI/ML models relies significantly on training data quality. Models are trained on labeled datasets with vulnerable and safe code samples, which are usually retrieved from platforms such as GitHub, SARD, or live-world exploit databases. Feature extraction techniques are used to map code to numerical representations—extracting such metrics as code complexity, function call graphs, control flow patterns, and comment semantics. Proper labeling of vulnerabilities allows supervised models to be trained on discriminative differences, while semisupervised and unsupervised models allow the identification of latent or unknown threats.

10.6.5 Evaluation metrics: accuracy, F1 Score, ROC–AUC

To ensure accuracy and consistency in vulnerability prediction models, they are assessed against a number of key metrics:

- Accuracy is total accuracy but deceptive in imbalanced data.
- F1 Score, as a harmonic mean of recall and precision, is a better approximation when false negatives or false positives are expensive.
- ROC–AUC (Receiver Operating Characteristic—Area Under Curve) is used to measure the model's ability for class discrimination with different sets of thresholds and hence is fitting in cases of binary classification in security scenarios.

These evaluations give a balanced perspective of model performance and inform subsequent optimization for use in production.

10.7 PREVENTION AND RESPONSE MECHANISMS

Since threats online are ever-evolving, vulnerability prediction alone would be inadequate. Sound security policy in Cloud DevOps pipelines must also include active prevention and wise response methodologies. AI and ML introduce a powerful driver to policy through providing dynamic, adaptive controls that continue to develop and mature in lockstep with the continually changing threat environment. ML and AI technologies give the capabilities to detect anomalies in real time, autonomous countermeasures, and learning continuously from historic threats.

10.7.1 Anomaly detection during build and deploy phases

Build and deployment is quite possibly the most crucial stage of CI/CD pipeline where code gets deployed and packaged into prod envs. Anomaly detection tools are supplemented by ML in order to monitor activity such as suspicious config updates, strange build artifacts, or deviating deployment patterns from baseline. These steps utilize techniques such as unsupervised learning (e.g., Isolation Forest and Autoencoders) to detect anomalies that may indicate a possible security incident or misconfiguration. Having anomaly detection built into these steps allows security teams to prevent malicious rollouts before production.

10.7.2 Automated response strategy using reinforcement learning

Reinforcement learning (RL) has been worth it in the development of response automation systems. In it, RL agents are trained to perform optimal actions—e.g., withdraw access, divert traffic, roll back releases, or quarantine containers—based on observed attack or deviation patterns. Over time, the model is trained based on a reward-penalty mechanism, hence optimizing efficiency because threats are dynamic in nature. The look-ahead feature optimizes mean time to response (MTTR) and avoids potential harm in high-speed DevOps environments.

10.7.3 Risk-based access control using ML profiling

Current access models use static permissions and roles that are less dynamic than responding to the dynamic cloud-native world. RBAC is ML-based behavior profiling-based risk-based access control, providing real-time estimates of user risk scores. The model will determine whether to allow, deny, or authenticate depending on user behavior, device fingerprint, log history, and geography. The dynamic methodology is immune to few to zero insider threats and credential misuse alternatives because it promises policies to react to user activity and surroundings.

10.7.4 Intermingling ML models within the CI/CD pipeline

AI/ML security controls are to be strongly ingrained within the CI/CD pipeline in such a manner that they can truly function. It is implemented within a series of checkpoints, starting from:

- Precommit hooks that use ML models to validate code prior to committing the code to the repository.
- Container scanning, where tools based on ML scan containers for known vulnerabilities, misconfigurations, or malware prior to deployment.
- IaC validation, where models validate configuration scripts (e.g., Terraform and Ansible) for policy noncompliance or insecure patterns. Such insertion of ML code blocks guarantees seamless, round-the-clock checking of code, dependencies, and deployment artifacts—greatly reducing human effort and making security at scale much stronger.

10.8 TOOLS AND PLATFORMS

The intersection of ML and AI with DevSecOps has brought in the development of various software and platforms, which have been created to enhance security across the DevOps development life cycle within Cloud. These provide vulnerability scan and compliance testing automation, along with injecting intelligence across various stages in CI/CD pipelines. From enterprise-based solutions to open-source solutions, the landscape is evolving at a rapid rate, and now there can be more scalable and long-lasting security architectures.

10.8.1 AI-based DevSecOps tools: Snyk, Aqua, Prisma Cloud

There are also new generations of AI-based DevSecOps tools that are capable of detecting and eliminating threats in real time.

- Snyk is an open-source vulnerability management specialized tool and is therefore inherently a part of developer workflows, providing AI-powered suggestions to patch security vulnerabilities in code, containers, and dependencies.
- Aqua Security employs ML models to detect behavioral anomalies in container workloads and enforces runtime defense.
- Palo Alto Networks' Prisma Cloud employs ML analytics for hybrid and multicloud end-to-end visibility, risk scoring, and threat protection. Such solutions enable robust integrations with CI/CD pipelines and cloud-native platforms to build highly scalable smart defense systems.

10.8.2 Open-source tools enhanced by ML: SonarQube, Clair, and Trivy

Certain open-source security tools are being enhanced by ML capabilities right now to provide smarter vulnerability scans.

- SonarQube, which has been used before for static code analysis, is now using ML-based models to detect fine-grained code smells, security hotspots, and patterns typical of suspected exploits.
- Clair, a scanner for container security, can be supplemented with AI modules to order threats based on contextual information and historical patterns of attacks.

Another widely adopted open-source scanner that has started utilizing AI to help with dependency scanning is Trivy, which helps teams tag and prioritize threats effectively later in build time. Extensibility and community creativity are a few of the reasons why such tools are used extensively.

10.8.3 Cloud-native services: AWS Macie, Azure Sentinel, Google Chronicle

Cloud vendors of large sizes have native AI-based services, which allow customers to build secure and compliant DevOps environments.

- AWS Macie applies ML to automatically categorize and protect sensitive information such as personally identifiable information within S3 buckets.
- Azure Sentinel is a cloud-native SIEM service that relies on ML and threat intelligence to detect advanced attacks on Azure and hybrid landscapes.
- Google Chronicle uses big-scale analysis and ML across security telemetry to allow security teams to quickly detect, hunt, and respond to threats. These services scale out-of-the-box, contain low config, and are entirely integrable with existing cloud infrastructure and therefore represent integral building blocks in AI-driven DevSecOps.

10.9 IMPLEMENTATION FRAMEWORK

AI/ML security integration into Cloud DevOps pipelines should have a well-thought-out and well-planned framework that can easily integrate into the rapid CI/CD pipeline. The architecture components and components involved in the deployment of ML models in DevOps operations to effectively detect and predict vulnerabilities are explained in the following section.

10.9.1 ML-Based Pipeline Security Architecture

A strong pipeline security system founded on ML starts with the delivery of intelligence at strategic points throughout the DevOps life cycle—namely, code commit, build, deployments, and operations. The system typically consists of a data ingest layer, ML model orchestration layer, and a mechanism for actionable feedback. Precommit hooks, container scan stages, and postdeployment monitoring integration points offer the nodes through which real-time feedback in the shape of security intelligence is fed back in. Architecture must support streaming as well as batch data analysis to be able to offer reactive and proactive threat detection.

10.9.2 Data sources: logs, code commits, infrastructure configs

The type of data sources and the quality of data sources are particularly important to ML-based security. Some of the most important inputs include

- Application logs and audit trails for behavior modeling and anomaly detection.
- Application source code commits for static vulnerability scans and analysis of code.
- IaC setups like Terraform or Kubernetes manifest to identify misconfigurations and unsafe defaults. Utilize the various datasets as a training set to offer good, contextual threat prediction models.

10.9.3 ML model integration in DevOps pipelines

Integration of ML models in DevOps pipelines calls for scale-out, nonintrusive, and lightweight tools and mechanisms. Models can be called by containerized microservices and called during CI/CD stages like premerge check or postbuild checks. Cloud-native functions like AWS Lambda or Azure Functions can also be employed for calling ML-based tests without clogging the pipeline. Model inference is best to be done in real time or near real time to enable developers and security teams to respond quickly.

10.9.4 Continuous learning loops and feedback systems

The feedback loop is likely the most important part of the model, which allows models to learn from the system. When new threats are identified or false positives are found, the system should be capable of taking such feedback and looping it back into model training cycles. This training cycle regularly updates the ML models to keep them current, accurate, and responsive to changing threat profiles. Active learning, web-based training, and retraining by schedule are often employed to keep model accuracy in alignment after some time has passed.

10.10 CASE STUDY

To demonstrate practical applications of AI/ML to DevOps security, the use of a ML model in forecasting CVEs of container images is explored in this study. To this end, the end-to-end model integration life cycle from data ingest through performance benchmarking in the CI/CD pipeline infrastructure is described.

10.10.1 Example: CVE prediction in container images with ML

Containerized applications would probably inherit vulnerabilities from base images, stale libraries, or insecure configurations. The purpose of the case study is to train and use a ML model to forecast the occurrence of CVEs on built features and metadata of container images. Features most important for examination could be OS type, installed packages, version history, Dockerfile patterns, and past CVE records.

10.10.2 Step-by-step application in a CI/CD pipeline

Step 1: Data retrieval and tagging: A score of container images are downloaded from Docker Hub, and other analogous CVE data relating to them are downloaded from the National Vulnerability Database, etc. All the images are tagged as vulnerable or not vulnerable on the basis of available CVEs.

Step 2: Feature Engineering: The metadata of base images, OS versions, package lists, and exposure ports are obtained. NLP operations tokenize Dockerfiles and identify malicious commands (e.g., RUN curl | bash). Feature vectors are constructed for ML training.

Step 3: Training and Model Selection: Various supervised learning models like Random Forest, SVMs, and Gradient Boosted Trees are trained and cross-validated. Deep learning models like LSTM are also experimented with for sequence-based Dockerfile parsing. They are evaluated on typical metrics like accuracy, precision, recall, and F1-score.

Step 4: Integration in CI/CD Pipeline: The model is bundled and implemented as a phase in the CI/CD pipeline. During building, every container image is scanned by the ML service to verify the probability of having a severe CVE. High-risk images are blocked from deployment and routed for manual checking.

Step 5: Reporting and Alerting: Scanner results are pushed into the DevSecOps dashboard, exposing vulnerable builds. DevOps and

Security teams are alerted and provided remediation recommendations, i.e., CVE patch links or more secure base images.

10.10.3 Analysis of performance and outcomes

The ML engine was found highly effective in uncovering potentially vulnerable images prior to actual CVE release, which carries predictive value. Key outcomes are as follows:

- **Model Accuracy**: 93%, F1-measure of 0.89
- **False Positive Rate**: 40% less than rule-based scanners
- **Deployment Latency**: Extremely low (<3 seconds per scan)
- **Improved Security Posture**: More than 60% reduction in postdeployment vulnerabilities

The integration also resulted in improved coordination between the developers and the security engineers, with proactive fix vs. reactive patches.

10.11 CHALLENGES AND LIMITATIONS

Though the incorporation of AI/ML technologies within cloud-based DevOps security pipelines has revealed unrealistic possibilities, the principal challenges and constraints remain the impediments to collaborative adoption and operational effectiveness. Inevitable detection of such limitations is the responsibility of the researchers, practitioners, and security architects who are involved in activities toward secure and reliable CI/CD environments.

10.11.1 Availability and quality of data

The most prominent issue with creating reliable AI/ML models for DevOps security is the quality of accessible training data. Security data can often have class imbalance (too few positive examples of attacks), noisy logs, inconsistent formatting, or limited public repositories. In addition, organizations also hesitate to share proprietary security data due to privacy, compliance, or intellectual property issues, which dissuades the development of generalized models. The lack of labeled and real-time data also dissuades supervised learning techniques and dissuades model resilience in real-world deployments.

10.11.2 Alert fatigue: false positives/negatives

AI/ML-powered security products, particularly those that run in high-speed DevOps environments, are prone to producing false positives or negatives, which makes developers and security teams lose faith in them. Too many too-frequent false alarms—the so-called alert fatigue—make teams feel overwhelmed by responses that are either late or, worse, missing the most important threats. False negatives, on the other hand, can allow actual threats to go below the radar and cause incursions. Reversing the tradeoff between sensitivity and specificity to achieve the optimum is still a problem, particularly with models running across numerous disparate environments with varying configurations.

10.11.3 Model explainability AI (XAI) challenges

Usage of advanced ML and deep learning models requires model explainability and interpretability challenges (e.g., LSTM and transformers). In cybersecurity, particularly in compliance scenarios, there needs to be some mechanism for security professionals to see the rationale behind each alert or categorization. Black-box models lack such explainability, which does not permit justification of automated action. Ineffective XAI tools intended for security-specific use continue to enhance stakeholders' trust and hinder incident response processes.

10.11.4 Overhead of integrating DevOps ecosystems

DevOps is dependent on speed, automation, and agility. Installing AI/ML-based layers of security—most importantly, those with high-data-preprocessing intensity, model training, or computationally expensive overhead—will interrupt build, test, and deploy operations. The majority of organizations will not include them since they do not desire the perceived latency overhead, resource usage, or ML infrastructure learnability overheads. Integrating the AI-based security products into CI/CD pipelines noninvasively continues to be a challenge.

10.12 BEST PRACTICES AND RECOMMENDATIONS

Implementation of AI/ML-enabled security control to cloud-native DevOps should be cross-functional and strategic in nature. As a means of supporting effectiveness, scalability, and resiliency, some best practices need to be enforced. These best practices strengthen security, besides unifying development, operations, and security teams under one vision of secure software delivery.

10.12.1 Shift-left early vulnerability prediction

Another key element of contemporary DevSecOps practice is the shift-left method, where security verification happens as early as possible during the software development life cycle. Prebuild time or code commit time, teams can forecast vulnerabilities with AI/ML models via static code analysis and NLP-based solutions. This is done earlier so that remediation cost is minimized and downstream vulnerability propagation is prevented. Early incorporation of prediction models enhances code hygiene and guarantees security as a collaborative effort from the beginning.

10.12.2 Ongoing verification of ML models

Security threats evolve rapidly, and so should the models that identify them. AI/ML pipelines should be continuously verified, retrained, and audited to stay accurate and current. Automated feedback loops allow models to adapt to evolving attack patterns, misclassifications, and false positives. Through A/B testing, concept drift detection, and model versioning, deployed models are ensured to keep getting better on new data without sacrificing precision and recall.

10.12.3 Machine learning pipelines securing (MLSecOps)

Similar to application pipelines, ML pipelines made up of data acquisition, preprocessing, training, and deployment are also vulnerable. MLSecOps best practices are model repository security, training data poisoning attacks protection, access control, and model provenance verification. Protect secure data in transit and at rest, and log training pipelines, so the AI layer itself is not a vector of attack.

10.12.4 DevSecOps culture establishment

Technology and tools cannot stop a pipeline unless an organizational culture is healthy. Cross-functional collaboration between security experts, developers, and operations teams is needed for the establishment of a DevSecOps culture. Frequent training in secure coding practice, threat model meetings, and awareness of tools powered by AI enhance the overall team potential. Security champions integrated into development teams and security-as-code practices can ensure a continued active security culture.

10.13 FUTURE TRENDS AND RESEARCH DIRECTIONS

As DevOps, cloud computing, and AI remain on the upgrade path, the prospect of AI/ML-based security in DevOps pipelines holds promises of thrilling developments. They do not simply target making systems smarter and secure, but increasingly, more transparent, private, and automated too. New technologies and paradigms define the next-gen secure DevOps environments.

10.13.1 Explainable AI for trust and transparency

Arguably, the biggest challenge in implementing AI/ML models for security is that they are not interpretable. Black-box models are strong and immune to present human-understandable reasons for their predictions, making security teams hesitant to implement them. XAI tries to fill the gap by providing transparent, explainable explanations of model choices. This strengthens trust, auditability, and regulatory compliance, especially in high-security domains such as finance, healthcare, and defense where it is as crucial to understand why an identified threat was found as the fact that it was found.

10.13.2 Federated learning for secure, privacy-aware training

Effective ML model training normally involves having enormous amounts of varied data, which could involve sensitive organizational or user data. Federated learning addresses this issue by allowing decentralized training on diverse nodes or organizations without sharing raw data. The method preserves data privacy while not sacrificing the value of collective intelligence. In DevOps security, federated learning is applied to construct global threat models between companies, enhancing predictive capability without compromising confidentiality.

10.13.3 AI-Augmented DevSecOps engineers (human–AI partnership)

Instead of human replacement, the future belongs to AI-augmented engineering, where clever systems augment security professionals with decision support, triaging alerts, and discovering emerging threats. Such human–AI collaborative models will assist DevSecOps engineers to concentrate on upper-level activities like strategic defense planning while AI agents take care of mundane tasks. Smart assistants may also provide security patches, automate reporting, and orchestrate remediation routines, changing teams' work patterns.

10.13.4 Automated security mitigation using self-healing systems

Self-healing systems are increasingly gaining popularity where systems sense and react to security violations automatically without any manual intervention. By using RL and closed-loop feedback mechanisms, upcoming systems will rollback vulnerable deployments, reconfigure incorrectly configured services, and isolate breached containers automatically. This model of automated resilience will revolutionize incident response by reducing mean time to recovery (MTTR) by orders of magnitude and downtime.

10.14 CONCLUSION

The convergence of AI, ML, cloud computing, and DevOps has transformed the world of modern software development and security. The current chapter explained how AI/ML technologies are increasingly being incorporated into DevOps pipelines to predict and prevent vulnerabilities, converting traditional security paradigms to intelligent, automated, and dynamic defense systems. We started with a general overview of Cloud DevOps pipeline basics and the particular security issues they raise at different points in the CI/CD process. The limitations of conventional, rule-based security controls—above all, their inability to match the pace and complexity of DevOps—highlight the need for more intelligent, more prescriptive security controls.

With close reading of the literature and analysis, we ventured into the application of ML techniques in static code analysis, anomaly detection, threat scoring, and threat intelligence in real time. From supervised classifiers to deep neural networks and NLP-based methods such as CodeBERT, their capabilities have shown fascinating outcomes in detecting patterns and vulnerabilities hidden in code and infrastructure settings. The utility of predictive and proactive security cannot be overemphasized. By identifying threats prior to exploitation and automating response mechanisms, organizations can minimize their attack surface and mean time to resolution. Additionally, the incorporation of AI into security functions guarantees protection not only responds but acts pre-emptively—shifting left and infusing every point in development and deployment with resilience. In the years to come, the vision is clear: creating intelligent, self-protecting DevOps environments that learn, adapt, and evolve continuously. Future breakthroughs in explainable AI, federated learning, human–AI collaboration, and autonomous response will continue to enhance the capacity to secure pipelines without compromising agility and innovation. As companies adopt these innovations, the symbiosis between AI and DevOps will play a key role in delivering scalable, sustainable, and dependable digital ecosystems.

REFERENCES

Arora, K., Joshi, G. P., Ragab, M., Rawa, M., Milyani, A. H., Mansour, R. F., & Yang, E. (2022). Bilateral contract for load frequency and renewable energy sources using advanced controller. *Computers, Materials & Continua*, *73*(2), 3165–3180. https://doi.org/10.32604/cmc.2022.026966

Arora, K., Kumar, A., Kamboj, V. K., Prashar, D., Jha, S., Shrestha, B., & Joshi, G. P. (2020). Optimization methodologies and testing on standard benchmark functions of load frequency control for interconnected multi-area power system in smart grids. *Mathematics*, *8*(6), 980. https://doi.org/10.3390/math8060980

Fu, M. (2023). AI for DevSecOps: a landscape and future opportunities. In: *Proceedings of the 2023 ACM Conference on Security and Privacy in Artificial Intelligence*, Copenhagen, Denmark, pp. 1–9.

Gugulotu, K. (2023). Integrating AI/ML into DevSecOps: strengthening security and compliance in cloud-native applications. *International Journal of Computer Engineering and Technology*, *15*(5), 107–115. https://doi.org/10.34218/IJCET.15.5.2023.107IAEME.

Liang, P., Wu, Y., & Xu, Z. (2024). Enhancing security in DevOps by integrating artificial intelligence and machine learning. *Journal of Theory and Practice of Engineering Science*, *4*(2), 31–37. https://doi.org/10.53469/jtpes.2024.04(02).05

Saleh, S. M., Sayem, I. M., Madhavji, N., & Steinbacher, J. (2024). Advancing software security and reliability in cloud platforms through AI-based anomaly detection. *arXiv preprintarXiv:2411.09200*. https://doi.org/10.48550/arXiv.2411.09200arXiv

Chapter 11

Design considerations for electric machines in IoT-based parking infrastructure

Pawandeep Kaur, Kumud Tiwari, and Sumit Sunil Shelar

11.1 INTRODUCTION

With the rapid evolution in society, the challenge of parking limited availability and providing a secure environment is an issue that plagues both drivers and the organization. As organizations continue to grow, the demand for secure entry and parking system escalates, exacerbating congestion, pollution, and frustration among commuters. Traditional entry and parking management systems, reliant on manual monitoring and static signage, struggle to keep pace with the dynamic needs of modern urban environments. Consequently, there is a pressing need for innovative solutions that optimize secure entry and parking system utilization, improve user experience, and reduce the environmental impact of urban mobility. To maintain privacy and security in society, which is the society's highest priority, this purpose work is specifically helpful for large organizations in keeping the area secure so that only registered, known individuals can enter through cars that close security system gaps.

One of the most crucial elements is security that must be handled in an organization for security reasons; therefore, the organization requires a system it can rely on to be secure. The combined need, which is significant from a security perspective, is included in this project research so that everyone in the organization will feel secure.

This study addresses these challenges by proposing an Internet of Things (IoT)-based smart secure entry and parking system, leveraging and integrating the Arduino UNO [1], NodeMCU, and GSM module to revolutionize Entry parking management in big organizations. The system integrates advanced technologies, including a passive radio frequency identification (RFID) scanner [1], to detect the authority of the user and allot parking slots and prioritize users on normal and VIP basis. By harnessing the power of IoT, the system provides real-time monitoring of parking availability [2], enabling a secure system by deducting charges from unauthorized users and allotting parking slots in a given time slot. The Arduino UNO and NodeMCU microcontroller serves as the central control unit, automating parking slot and time allocation and enhancing system efficiency.

 DOI: 10.1201/9781003641582-11

The proposed secure entry and parking system is particularly relevant in the context of educational and big organizations, such as college campuses during technical fests, where parking demand fluctuates significantly. By allocating parking slots based on real-time availability and user priority [3], the system optimizes parking space utilization and minimizes the time spent searching for parking. Additionally, the integration of components such as Arduino, liquid crystal display (LCD) with I2C module, servo motor, NodeMCU, GSM module, RFID scanner, and breadboard enhances the functionality and effectiveness of the system, further improving user experience.

In summary, this study presents a novel approach to address the secure entry and parking challenges in crowded areas or organizations through the implementation of an IoT-based smart parking system (SPS) using the Arduino UNO, NodeMCU, and GSM module. By providing real-time monitoring of parking availability and prioritizing users based on demand, the proposed system offers a promising solution to optimize parking space utilization, reduce congestion, and enhance the overall urban mobility experience.

11.2 LITERATURE SURVEY

The literature surrounding SPS and IoT-based solutions reveals a growing interest in addressing the challenges of urban parking management. Researchers have explored various approaches and technologies to improve parking space utilization, enhance user experience, and reduce the environmental impact of urban mobility. According to previous research, we did find a few existing promising systems available. All of them were made using different technologies and were focused on different purposes.

The author of [1] paper proposed an Arduino-based SPS, with Fee Management Using Arduino. To this day, people are still using manual parking systems that consist of many problems, for example, looking for free space in a parking lot without any prior knowledge if the parking lot is full or not, which leads to wastage of time and fuel. Safety of vehicles is also an addressable issue. Keeping this in mind, we are proposing an automobile parking system. Through this system, drivers will be able to know if there are vacant parking slots or not, along with the vacant parking slot number(s) prior to entering the parking lot via monitor screen(s) at the entry gate(s). Also only authorized persons shall be given entry access to the parking lots via an RFID tag depending on the availability of parking space. Furthermore, we also introduced a parking fee management system, which is also payable using the same RFID tag that acts as a rechargeable parking credit card.

In another paper [2], the author proposed a method that enables the user to find the nearest parking area and gives availability of parking slots in that respective parking area. And it mainly focuses on reducing the time in finding the parking lots and it avoids unnecessary traveling through filled parking lots in a parking area. Thus, it reduces fuel consumption which in turn reduces carbon footprints in an atmosphere. The system is a Raspberry pi-based parking sensor that contains a pi camera to detect the empty parking spaces and sends this data to server, and this stored data is accessed by users.

In paper [3], the author designed a smart car parking system. When a person with disability tries to park a vehicle in a city having heavy traffic, he faces several problems in such a situation to search and park the vehicle difficult. The disabled person can park the vehicle in a specially designed location. The intelligent parking finds the best available parking, minimizing the cost. Parking and unparking can be done with the help of an android application. The user has to install the android application on his/her smartphone devices, which would have the facility to prebook the parking slot.

In the following paper [4], the author's aim of this project is to carmate the car and car parking as well. A miniature model of a Carmated car parking system can regulate and manage the number of cars that can be parked in a given space at any given time based on the availability of the parking slot. Carmated parking is a method of parking and existing cars using a sensing device.

In paper [5], the author explains the architecture and design of Arduino-based car parking system. Authorization of driver or user is the basic rule used to park a vehicle in a parking place. An authorization card will be given to each user, which carries the vehicle number or other details. If the user is authorized and space is available in the parking lot, then the parking gate will open, and the user is allowed to park the vehicle in a parking place else the user is not allowed even the user is an authorized person. If a car is allowed to park, then a mobile notification will be sent to the user about parking. The parking issue in urban areas also provides security to a vehicle and an unauthorized user is not allowed to enter a parking place. It helps park vehicles in a multifloor car park also as it will display which floor has free space.

In paper [6], the authors aim to present an intelligent system for parking space detection based on an image processing technique that captures and processes the brown-rounded image drawn at a parking lot and produce the information of the empty car parking spaces. It will be displayed at the display unit that consists of seven segments in real time. The seven segments display shows the number of current available parking lots in the parking area. This proposed system has been developed in software and hardware platform.

In paper [7], the authors proposed an Intelligent Car Parking Management and Security System that addresses parking congestion and vehicle theft. The system automates vehicle entry and exit using microcontrollers, infrared (IR) sensors, and RFID technology. It reduces the time required to find an available parking slot by displaying the status of each slot on an LCD screen. The use of RFID tags at entry and exit gates provides enhanced security, as only authorized vehicles can enter or leave. The system's automation reduces human intervention and improves parking management efficiency, making it faster and more secure.

In paper [8], the authors propose a method that helps users locate the nearest parking area and check the availability of parking slots. This approach minimizes the time spent in searching for parking, reduces unnecessary travel through full lots, and subsequently lowers fuel consumption and carbon footprints. The system utilizes a Raspberry pi-based parking sensor equipped with a pi camera to detect empty spaces, sending this data to a server for user access.

In paper [9], Bhatt et al. propose a Car Parking Automation and Premises Security System aimed at addressing the growing parking challenges in India. The system utilizes a Vertically Automated Rotary Parking System, requirings a minimal human interaction, allowing drivers to drop off their cars at the entrance for automated parking. This method optimizes space, enhances convenience, and improves safety while reducing environmental impact. Additionally, the project incorporates an advanced security system featuring number plate recognition and fingerprint scanning, ensuring controlled access and enhancing safety for residents. Overall, this dual approach offers a more efficient and secure parking solution for urban areas.

In paper [10], Fahim et al. conduct a comprehensive review of SPSs addressing the growing parking challenges in modern cities. The study analyzes various SPS technologies, including sensors, networking techniques, user interfaces, and computational approaches. The authors highlight the significant issue of parking space scarcity, noting that around 30% of urban vehicles search for parking, causing traffic congestion and air pollution. They categorize SPSs into 12 groups based on methods like IoT, machine learning, and computer vision and discuss the pros and cons of various sensors used in these systems. By comparing existing SPS solutions, the paper aims to assist researchers, designers, and policymakers in identifying effective systems and understanding current trends in smart parking technology.

In paper [11], Arora et al. developed an SPS prototype using Arduino and sensors to address urban parking challenges. The system incorporates ultrasonic sensors for obstacle detection and a Passive Infrared (PIR) sensor to identify car movement in parking spaces. This setup guides drivers to available spots, reducing traffic congestion. The system features alarms for safe parking and real-time data on space availability. The design and software implementation are detailed, emphasizing the prototype's potential to enhance urban mobility and parking management through improved safety and efficiency.

In paper [12], Balmiki et al. present an IoT-based Smart Vehicle Parking System designed to address the increasing traffic and parking challenges in urban areas. The system employs RFID technology to automate vehicle identification and parking slot management, allowing custodians to easily find available spaces. Unlike existing systems, it integrates real-time availability checks and automates the payment process to reduce fraud and enhance user convenience. The methodology includes a countdown mechanism for vehicle entries and exits, with RFID cards facilitating seamless access and fee calculation based on time spent. The proposed system aims to minimize fuel consumption and carbon emissions while improving overall efficiency in parking management. Future work involves expanding the system to contribute to smart city initiatives and integrating it with broader traffic management frameworks.

In paper [13] by Elsonbaty and Shams, a Smart Parking Management System (SPMS) was developed to address the growing parking challenges in urban areas. This system leverages Arduino components and IoT technology, enabling users to check available parking spaces and make reservations through a mobile application. Using IR sensors, the system detects whether parking slots are occupied, and this data is transmitted via Wi-Fi to a server accessible via the app. The SPMS aims to reduce the time spent searching for parking and decrease fuel consumption, ultimately improving urban mobility. Users can receive real-time updates about parking availability and manage their reservations conveniently through the app.

In paper [14], the authors explore automated underground car parking systems to address the growing parking challenges in urban areas. As the number of vehicles increases, finding parking becomes more difficult, especially in congested cities. Conventional underground parking garages have specific design requirements, such as unobstructed access, adequate head clearance, and improved lighting for safety. The automated systems aim to enhance efficiency, reduce environmental impact, and optimize space usage. The report emphasizes the importance of innovative solutions for effective urban planning and managing parking demand.

In paper [15], the authors designed a smart car parking system aimed at addressing the challenges of urban parking. As the number of vehicles increases, drivers often struggle to find available parking spaces, leading to congestion and frustration. The proposed system utilizes Arduino microcontrollers, IR sensors, and servo motors to provide real-time information on parking space availability. It guides drivers to vacant spots, helping optimize parking utilization and reduce traffic issues. The system is adaptable to various parking environments and is designed to be cost-effective. With an intuitive user interface, drivers can access real-time parking data, including the option to reserve spaces in advance. By streamlining the parking process, this smart car parking system significantly enhances the overall parking experience in busy urban areas.

In paper [16], the authors present a wireless sensor network solution for car-park management. The system is designed to monitor parking space availability and provide real-time information to users and facility administrators. It utilizes DSYS25z sensing nodes, which operate on a custom communication protocol due to challenges with existing protocols. Key findings include unexpected communication reliability patterns: the presence of vehicles can degrade signal quality, and link reliability varies over time but not space. The system architecture is tiered, with sensing, data forwarding, and data processing layers. A server collects data from the sensors and updates a MySQL database, allowing users to access a web application that displays parking availability on a map. This user-friendly interface supports administrative functions and reporting on occupancy trends, enhancing overall car-park management efficiency.

Paper [17] introduces a scalable algorithm for disseminating spatiotemporal traffic information, such as parking space availability, utilizing vehicular ad hoc networks (VANET) built on the IEEE 802.11 Wireless LAN standard. The algorithm relies on periodic broadcasts to share information while reducing redundancy through application layer evaluation and data aggregation. Given the dynamic nature of parking information, the algorithm incorporates techniques that consider the local relevance and age of the data, thereby limiting its spatial distribution. Simulations based on a realistic model of a German city, featuring up to 10,000 vehicles, demonstrate that a decentralized parking information system can operate effectively with just 5% of vehicles equipped with Wireless LAN capabilities, without requiring any message relaying infrastructure. This characteristic makes this nonsafety driver assistance application particularly appealing during the initial deployment phase of VANET.

Overall, the literature survey highlights the growing interest in IoT-based SPSs and their potential to address the challenges of urban parking management. By leveraging IoT technology, machine learning algorithms, and Arduino microcontrollers, researchers and industry stakeholders are working toward creating more efficient, user-friendly, and sustainable parking solutions for crowded urban environments.

11.3 METHODOLOGY OF THE PROPOSED SYSTEM AND DESIGN

11.3.1 System architecture and schematic circuit

In this proposed system, Figure 11.1 shows the architecture for VIP users, which depicts the integration of sensors with NodeMCU function as a microcontroller and how communication takes place between NodeMCU and RFID. The sensors include an IR sensor for vehicle detection for VIP users, RFID for

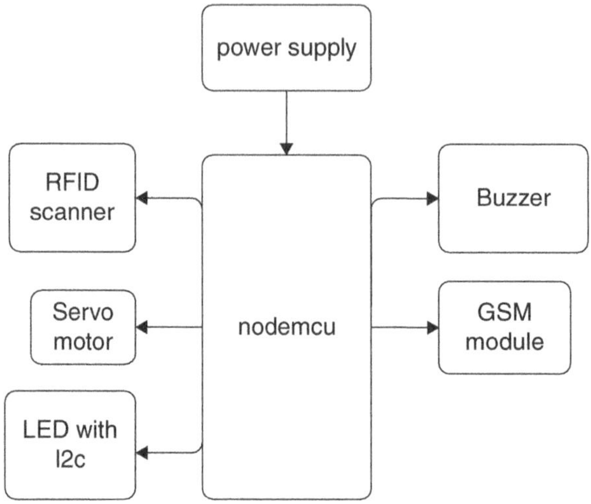

Figure 11.1 Proposed architecture of parking and entry system for VIP users.

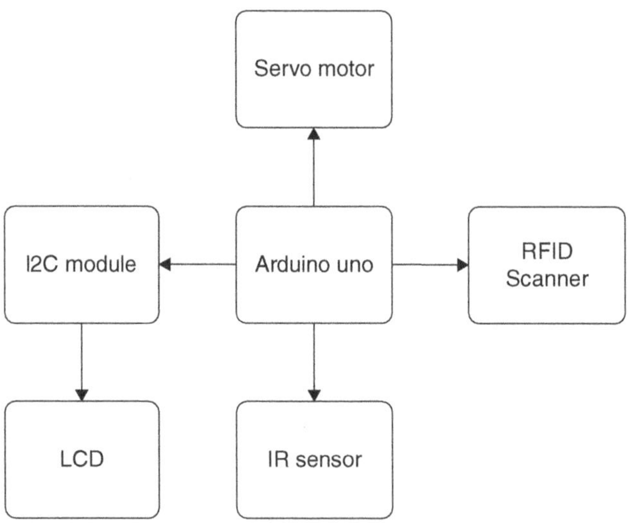

Figure 11.2 Proposed architecture of the parking system with Arduino.

card scanning, Servo motor for barrier, LCD for displaying information, and NodeMCU with the integration of GSM for alert message communication.

Figure 11.2 shows the architecture for VIP users, which depicts the integration of sensors with Arduino function as a microcontroller and how communication takes place between NodeMCU and RFID. The sensors include an IR sensor for vehicle detection for Normal users, RFID for card scanning, Servo motor for barrier, and LCD for displaying information.

11.3.2 Flow-graph

In the following sections, flow graphs are shown for NodeMCU esp8266, which is used for VIP lane as the car goes in the VIP lane and the VIP card scanned on an RFID scanner if the card is a VIP authorized card then the servo motor barrier gets open and a reserved slot allotted to the VIP user and if the scanned card is not a VIP authorized one then the security alert will be forwarded to the security office and buzzer will be on.

Two flow graphs are shown for the Arduino normal lane, which is for normal user. For example, if the faculty member or office person car goes in the normal lane and the normal authorized card scanned on RFID the barrier will be opened and the parking slot will be allotted and if the unauthorized user comes in the reason of entry entered manually by security office and a temporary card will be allotted for a selected time slot and the parking slot will be allotted and charges will be deducted from the user (Figure 11.3; Table 11.1).

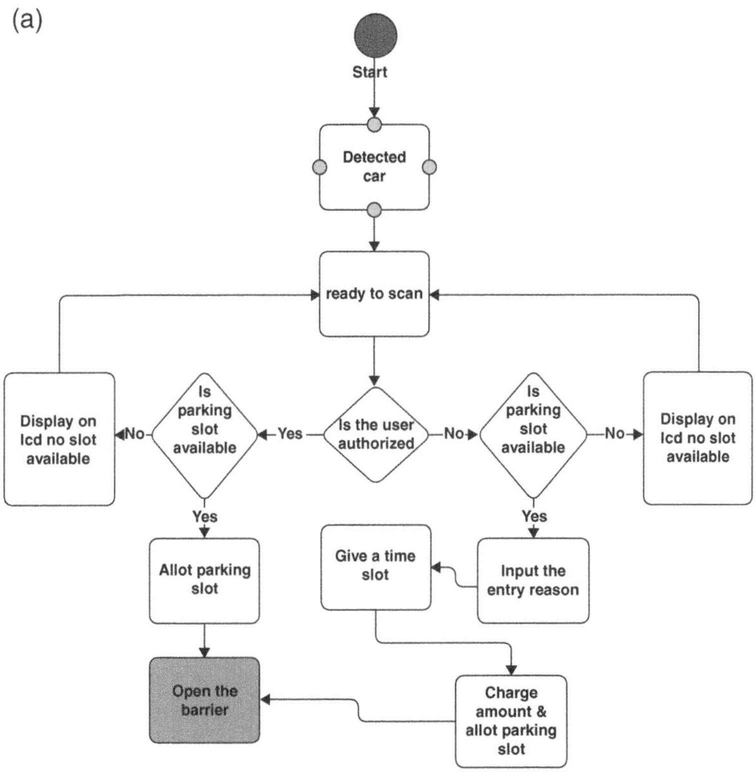

Figure 11.3 (a) Process flowchart for a normal user. (b) Process flowchart for the VIP lane.

(Continued)

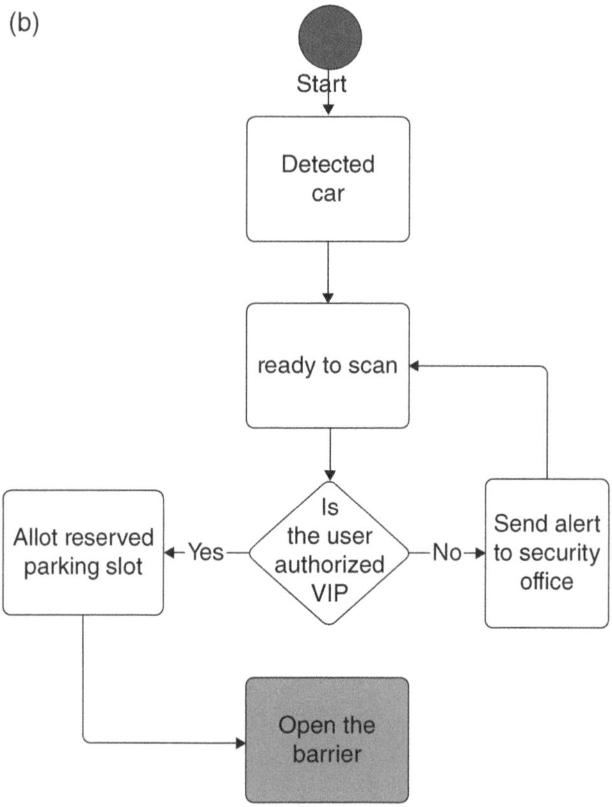

Figure 11.3 (Continued) (a) Process flowchart for a normal user. (b) Process flowchart for the VIP lane.

Table 11.1 Component required for implementing the proposed model

Serial no.	*Equipment*	*Quantity*
1	Arduino UNO R	1
2	IR sensor	1
3	Bread board	1
4	LCD	1
5	I2C module	1
6	Servo motor	1
7	RFID module	1
8	NodeMCU ESP8266	1
9	GSM module	1
10	Buzzer	1
11	Led	1
12	RFID card	3

11.3.3 Components

11.3.3.1 IR sensor

An IR sensor is an electronic device that detects and measures IR radiation. IR radiation is a type of electromagnetic radiation with a longer wavelength than visible light but shorter than radio waves. It's invisible to the human eye, but all objects with a temperature above absolute zero emit IR radiation (Figure 11.4).

11.3.3.2 Arduino

Arduino is an open-source electronics platform with a microcontroller that interacts with various electronic components [18]. It features digital and analog input/output pins, with the Arduino UNO being the most common. Programming is done using the Arduino IDE, a C/C++ programming language. Arduino's open-source nature allows for easy modification and building upon, leading to a large ecosystem of compatible boards and a vast community of developers (Figure 11.5).

11.3.3.3 NodeMCU

NodeMCU is an open-source development board and firmware for IoT projects, based on the ESP8266 Wi-Fi module. It offers a low-cost, powerful microcontroller with built-in Wi-Fi capabilities, making it ideal for projects requiring wireless connectivity. NodeMCU features GPIO pins for interacting with sensors and actuators, integrated Wi-Fi for direct connection to a Wi-Fi network, compatibility with LUA and Arduino IDE, and power and USB connections for serial communication. Its compact size, affordable cost, and ease of use make it an ideal choice for IoT projects (Figure 11.6).

Figure 11.4 Infrared (IR) sensor.

Figure 11.5 Microcontroller board: Arduino UNO.

Figure 11.6 ESP8266 Wi-Fi module.

11.3.3.4 LCD

LCD, which is a flat panel, is made up of liquid crystal in its primary form of operation. Arduino-based projects often display information on an LCD panel, with the 16×2 LCD being the most used model. These displays can be programmed and interfaced using the Liquid Crystal library in the Arduino

IDE [19]. The display uses multiple digital pins for data transfer, but an I2C module can simplify wiring and reduce the number of connections to just two (SDA and SCL). The LCD can display real-time data output, including sensor readings and system status (Figure 11.7).

11.3.3.5 RFID module

An RFID module is a tool used in electronics projects to read and write data to RFID tags. The RC522 RFID module, operating at 13.56 MHz, is commonly used in Arduino applications. The tag is powered by a signal from the module, allowing communication and returning its unique ID. The module consists of a microcontroller for data processing and an antenna for transmitting and receiving radio signals. RFID tags can be active or passive. RFID is the most important component in a view of security (Figure 11.8).

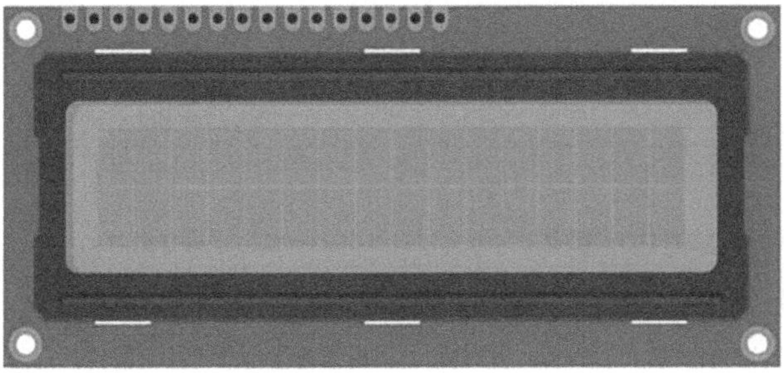

Figure 11.7 LCD (liquid crystal display).

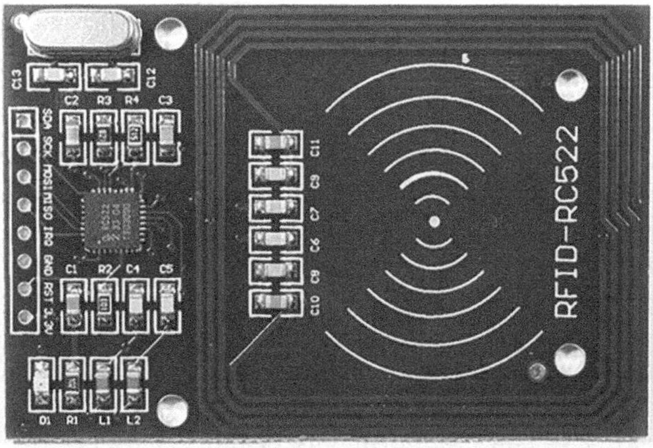

Figure 11.8 Radio frequency identification (RFID) module.

Table 11.2 Comparison table of Arduino and node MCU

Feature	*Arduino*	*Node MCU*
Microcontroller	At mega series	ESP8266
Connectivity	No built-in Wi-Fi	Built-in Wi-Fi (ESP8266)
Power consumption	Low power consumption	High power consumption due to Wi-Fi module
Development environment	Arduino IDE	Arduino IDE (with ESP8266 board support)
Programming language	C/C++	C/C++
Cost	More affordable	Slightly more expensive due to Wi-Fi module
Typical use cases	Robotics, IoT devices without Wi-Fi, prototyping	IoT devices requiring Wi-Fi connectivity, smart home applications
Key advantages	Simplicity, affordability, wide range modules	Built-in Wi-Fi, easy integration with cloud services

11.3.3.6 Comparison table on microcontroller boards

See Table 11.2.

11.3.4 Overview

The normal lane structure is created using an Arduino, along with an IR sensor, servo motor, and an RFID card module (Figure 11.9).

Here, the figure portrays a practical illustration of the system here; the LCD monitor at the entry can be seen which shows information about the parking lot such as free parking slot numbers [20]. When the VIP vehicle enters, it displays the message as in Figure 11.9b and when the normal unauthorized vehicle enters after allotting time slot it displays the message as in Figure 11.9e the charges get deducted and the slot allotted to the user and when the unauthorized car enters in the VIP lane it displays a message as in Figure 11.9f, the buzzer gets on and the alert will be sent to the security office as in Figure 11.9g.

11.4 RESULTS AND DISCUSSION

1. There are two lanes for normal users and VIP users.
2. It guarantees snappy and computerized parking and simple recovery of vehicles.

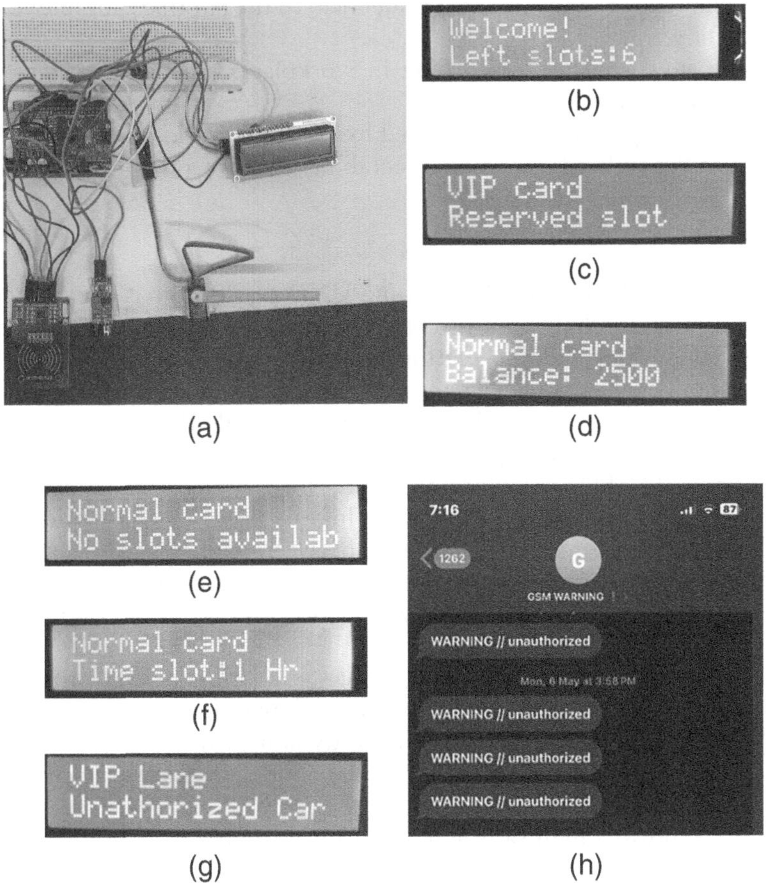

Figure 11.9 (a) Hardware representation. (b) At the start of the system, the first message shown on the display of the left parking slots when a normal car arrives. (c) When a VIP vehicle arrives, the system displays a message confirming the individual's identity and sends the notification. (d) When a regular organization-registered vehicle arrives and the card is successfully authorized, the system displays the remaining card balance after deducting the parking fee. (e) When normal car arrives but no slot is left for the car, this message is shown on the screen. (f) Time slot allotted to the normal registered or unregistered car as per the need of the person. (g) When a normal registered car or unregistered car arrives though the VIP lane, an alarm alert will be sent to the security person to check on the situation. (h) The sample of the message sent to the security office that an unauthorized car is in the VIP lane.

3. Up to six cars can be effectively and securely parked in the outlined model.
4. Reserved space for VIP personnel.
5. Security guarded with an unauthorized access alert.

6. The surface space required is identical to the parking spot of two cars as it were.
7. Most reasonable for parking in workplaces, shopping centers, and comparable spots.
8. Low support levels are required by the framework.
9. Sensors utilized have high affectability and are anything but difficult to deal with.
10. Security of vehicle.

The IoT-based SPS, using the Arduino UNO platform, has been tested to address parking challenges in crowded urban environments. SPS based on Nedemcu and Arduino platforms is aimed at addressing secure and efficient parking management and entry system in educational institutions and other organizations. The system integrates RFID scanners to identify user and employs a dual-priority scheduling mechanism: First Come First Serve (FCFS) for regular users and priority-based scheduling for VIP users [21]. Additionally, RFID technology is used for charge deduction and user identification, differentiating between regular and VIP users. By leveraging IoT technology, the system provides real-time monitoring of parking space availability, optimizing space utilization, and minimizing user search time.

The system optimized parking space utilization by allocating slots based on real-time availability and user demand, reducing congestion in parking lots. The system's ability to dynamically adjust parking allocations based on changing demand contributed to more efficient use of parking spaces [22].

Users expressed high satisfaction with the secured system, highlighting its effectiveness in reducing parking search time, enhancing user experience, and improving urban mobility. User feedback provided valuable insights for further system improvements, ensuring continuous optimization and adaptation to evolving user needs (Table 11.3).

11.5 CONCLUSION

The system optimized the utilization of parking spaces by assigning slots according to real-time availability and user demand, effectively reducing congestion in parking lots. Its ability to dynamically adjust parking allocations in response to fluctuating demand led to a more efficient use of parking resources.

Users reported high satisfaction with the secure system, noting its success in reducing the time spent searching for parking, improving their overall experience, and enhancing urban mobility. By utilizing PIR sensors to detect parking availability, the system provided real-time updates, enabling users to quickly and easily find vacant spots. The integration of LCDs with

Table 11.3 Comparison of the performance of an existing model with the proposed model

References	*Normal authorized car slot allocation*	*Normal unauthorized car time & slot allocation*	*VIP reserved slot allocation*	*VIP unauthorized entry alert*	*Unauthorized car charge deduction*	*Authorized vehicle no cash deduction*
[1]	Yes	–	–	–	Yes	–
[2]	Yes	–	–	–	–	–
[3]	Yes	Yes	–	–	–	–
[4]	Yes	Yes	–	–	–	–
[5]	Yes	Yes	–	–	–	–
[6]	Yes	–	–	–	–	–
[7]	Yes	–	–	–	–	–
[8]	Yes	yes	–	–	–	–
[9]	Yes	–	–	–	–	–
[10]	Yes	–	–	–	–	–
[11]	Yes	–	–	–	–	–
[12]	Yes	–	–	–	–	–
[13]	Yes	Yes	–	–	–	–
[14]	Yes	Yes	–	–	–	–
[15]	yes	Yes	–	–	–	–
Proposed work	Yes	Yes	Yes	Yes	Yes	Yes

I2C modules ensured that parking information was communicated clearly, further boosting user satisfaction and interaction.

The FCFS scheduling algorithm, implemented to allocate parking slots based on arrival time, promoted fairness and efficiency in the process. This transparency increased user confidence in the system and contributed to their overall satisfaction.

Compared to existing systems, this solution stands out for its responsiveness, reliability, and capacity to dynamically adjust to real-time changes in demand. Users consistently reported decreased search times and enhanced convenience, creating a better overall parking experience.

In summary, the IoT-based SPS marks a significant improvement in parking management, offering a scalable, sustainable approach to managing parking challenges in densely populated cities. Ongoing monitoring, evaluation, and updates will be critical to maintaining its effectiveness and relevance in ever-evolving urban settings.

To support larger parking facilities and diverse urban environments, the system's scalability should be enhanced by incorporating advanced networking capabilities. This will enable seamless communication between multiple parking lots and centralized management systems, allowing for more efficient coordination and resource allocation.

The system can be further improved by integrating with broader smart city infrastructure. By connecting with traffic management systems, public transportation networks, and urban planning efforts, the SPS can contribute to a more comprehensive approach to urban mobility, improving both traffic flow and accessibility.

Leveraging advanced data analytics can provide deeper insights into parking demand, user behavior, and overall system performance. Predictive modeling and proactive management strategies will help optimize parking space usage, increase operational efficiency, and enhance user experience by anticipating and responding to demand patterns.

REFERENCES

1. Kabir, A. Z. M. T., Rahman, M. M., & Hasan, M. R. (2019). Automated parking system with fee management using Arduino. In: *Proceedings of the 10th International Conference on Computing, Communication and Networking Technologies (ICCCNT 2019)*, Kanpur, India, pp. 1–6. IEEE.
2. Basavaraju, S. R. (2015). Automatic smart parking system using Internet of Things (IoT). *International Journal of Scientific and Research Publications*, *5*(12), 629–632.
3. Geng, Y., & Cassandras, C. G. (2011). A new "smart parking" system based on optimal resource allocation and reservations. In: *Proceedings of the 14th International IEEE Conference on Intelligent Transportation Systems (ITSC 2011)*, pp. 979–984. https://doi.org/10.1109/ITSC.2011.6082832.

4. Bonde, D. J., Shende, R. S., Gaikwad, K. S., Kedari, A. S., & Bhokre, A. U. (2014). Automated car parking system commanded by Android application. In: *Proceedings of the 2014 International Conference on Computer Communication and Informatics (ICCCI)*, 3–5 January 2014, Coimbatore, India. https://doi.org/10.1109/ICCCI.2014.6921729.
5. Chaudhary, H., Bansal, P., & Valarmathi, B. (2017). *Advanced car parking system using Arduino. In: 2017 4th International Conference on Advanced Computing and Communication Systems (ICACCS)*, Coimbatore, India (pp. 1–5). IEEE. https://doi.org/10.1109/ICACCS.2017.8014701.
6. Yusnita, R., Norbaya, F., & Basharuddin, N. (2012). Intelligent parking space detection system based on image processing. *International Journal of Innovation, Management and Technology*, *3*(3), 232–235. https://doi.org/10.7763/IJIMT.2012.V3.228.
7. Priyanka, K., Sirisha, G., Vijaya Lakshmi, K. N. V. S., & Sujatha, Y. (2017). Intelligent car parking management and security system. *International Journal for Scientific Research and Development*, *5*(3), 1612–1615.
8. Yan, G., Olariu, S., Weigle, M. C., & Abuelela, M. (2008). SmartParking: A secure and intelligent parking system using NOTICE. In: *Proceedings of the 2008 IEEE Intelligent Transportation Systems Conference (ITSC)*, Beijing, China (pp. 569–574). IEEE. https://doi.org/10.1109/ITSC.2008.4732675
9. Bhatt, B., Sharma, H., Arora, K., Joshi, G. P., & Shrestha, B. (2023). Levy flight-based improved grey wolf optimization: A solution for various engineering problems. *Mathematics*, *11*(7), 1745. https://doi.org/10.3390/math11071745.
10. Fahim, A., Hasan, M., & Chowdhury, M. A. (2021). Smart parking systems: Comprehensive review based on various aspects. *Heliyon*, *7*(5), e07050. https://doi.org/10.1016/j.heliyon.2021.e07050.
11. Prakash, G., Logapriya, P., & Sowmiya, A. (2024). Smart parking system using Arduino and sensors. *Naturalista Campano*, *28*(1), 2903–2911. https://doi.org/10.17605/OSF.IO/6N3E5.
12. Balmiki, D., Singhal, M., Singh, A., & Tyagi, D. (2023). A review on smart vehicle parking system. *International Journal of Scientific Research & Management Studies*, *4*(6), 120–123.
13. Elsonbaty, A. A., & Shams, M. Y. (2020). The smart parking management system. *International Journal of Computer Science & Information Technology*, *12*(4), 55–60. https://doi.org/10.5121/ijcsit.2020.12405.
14. Shaikh, A. M., Bergsma, J., & Thao, D. T. P. (2010). *Automated Underground Car Parking Areas*. Delft University of Technology. https://ocw.tudelft.nl/wp-content/uploads/UUS_Automated_Underground_Car_Parking_Areas.pdf.
15. Arora, K., Joshi, G. P., Ragab, M., Rawa, M., Milyani, A. H., Mansour, R. F., & Yang, E. (2022). Bilateral contract for load frequency and renewable energy sources using advanced controller. *Computers, Materials & Continua*, *73*(2), 3165–3180. https://doi.org/10.32604/cmc.2022.026966.
16. Chinrungrueng, J., Dumnin, S., & Pongthornseri, R. (2011). iParking: A parking management framework. In: *Proceedings of the 11th International IEEE Conference on Intelligent Transportation Systems (ITST)*, St. Petersburg, FL, USA (pp. 63–68). IEEE.

17. Mathur, S., Kaul, S., Gruteser, M., & Trappe, W. (2009). ParkNet: A mobile sensor network for harvesting real-time vehicular parking information. In: *Proceedings of the MobiHoc S3 Workshop on MobiHoc S3*, ACM, New York, NY, USA (pp. 25–28). https://doi.org/10.1145/1540358.1540367.
18. Wolff, J., Heuer, T., & Gao, H. (2006). Parking monitor system based on magnetic field sensors. In: *Proceedings of the IEEE Intelligent Transportation Systems Conference (ITSC)*, Toronto, Canada (pp. 1275–1279) IEEE. https://doi.org/10.1109/ITSC.2006.1707398.
19. Propst, J. A., Poole, K. M., & Hallstrom, J. O. (2012). An embedded sensing approach to monitoring parking lot occupancy. In: *Proceedings of the 50th Annual Southeast Regional Conference (ACM-SE '12)*, Tuscaloosa, AL (pp. 309–314) ACM. https://doi.org/10.1145/2184512.2184584.
20. Arora, K., Kumar, A., Kamboj, V. K., Prashar, D., Jha, S., Shrestha, B., & Joshi, G. P. (2020). Optimization methodologies and testing on standard benchmark functions of load frequency control for interconnected multi-area power system in smart grids. *Mathematics*, *8*(6), 980. https://doi.org/10.3390/math8060980.
21. Benson, J. P., O'Donovan, T., O'Sullivan, P., Roedig, U., & Sreenan, C. (2006). Car-park management using wireless sensor networks. In *Proceedings of the 3rd ACM International Workshop on Vehicular Ad Hoc Networks (VANET '06)*, Los Angeles, CA, USA (pp. 30–39) ACM. https://doi.org/10.1145/1161064.1161070.
22. Caliskan, M., Graupner, D., & Mauve, M. (2006). Decentralized discovery of free parking places. In: *Proceedings of the 3rd ACM International Workshop on Vehicular Ad Hoc Networks (VANET '06)*, Los Angeles, CA, USA (pp. 30–39). https://doi.org/10.1145/1161064.1161070.

Chapter 12

Integrating Industry 5.0 principles into electrical machine design

Exploring opportunities and challenges

Pankaj Kumar Keshri

12.1 INTRODUCTION

After 10 years of providing Industry 4.0 to address the industry's deficiencies, the time has come for Industry 5.0. Industry 4.0 is limited since smart factories are raising corporate productivity. Industry 5.0, which emphasizes the integration of cutting-edge technologies with human-centric approaches, marks a revolutionary transformation in the manufacturing environment [1–3]. This new paradigm aims to improve human–machine collaboration while balancing automation and artificial intelligence (AI) with ecological practices. Industries must manage worker adaptation, ethical issues, and technology integration while they take advantage of the opportunities brought forth by greater customization, efficiency, and resilience. To solve these issues and find creative solutions that maximize Industry 5.0's potential while maintaining environmental sustainability and social responsibility, future research areas will be essential [4,5].

12.2 INDUSTRY 5.0 TECHNOLOGIES

Cloud computing, blockchain, big data analytics, the Internet of Things (IoT), and 6G networks are some of the enabling technologies for Industry 5.0, as represented in Figure 12.1.

12.2.1 Cloud computing

Industry 5.0, which emphasizes the collaboration of humans and cutting-edge technologies to develop more customized and effective production processes, heavily relies on cloud computing. How Industry 5.0 technologies are supported by cloud computing is described below:

a. **Data Integration and Management**: Cloud computing offers the framework for integrating and managing the massive volumes of data produced by Industry 5.0 technologies, including robotics, sensors,

DOI: 10.1201/9781003641582-12

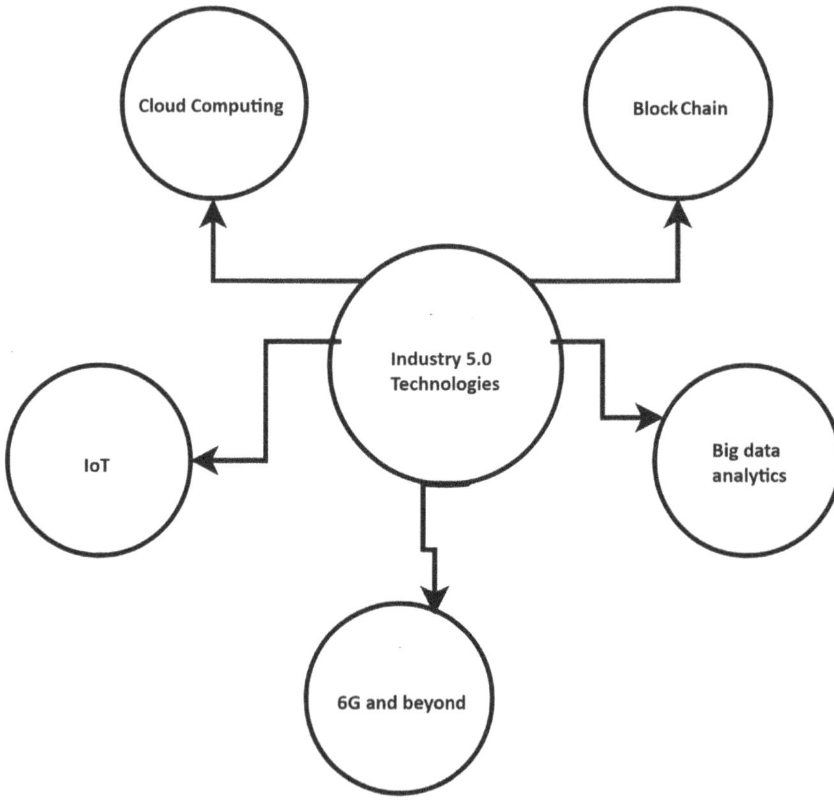

Figure 12.1 Industry 5.0 technologies.

and IoT devices. Making educated decisions and streamlining procedures depend on the ability to store data centrally and interpret it in real time.

b. **Scalability**: The cloud provides adaptive resources that can be changed in response to demand. For Industry 5.0 applications, where workloads may fluctuate owing to shifting production requirements or adjustments in data volumes, this flexibility is essential.
c. **Collaboration and Connectivity**: Cloud platforms provide smooth communication and cooperation across various teams and departments as well as between humans and robots. A common feature of Industry 5.0 is the combination of AI- and human-driven operations; cloud computing facilitates data exchange and easy communication between these entities.
d. **Advanced Analytics and AI**: Access to strong analytics tools and AI services that can analyze large, complicated datasets and produce insights is made possible by cloud computing. In Industry 5.0

scenarios, these tools are critical for quality control, predictive maintenance, and optimization.

e. **Cost-Effectiveness**: Businesses can lower the capital costs related to maintaining on-premises technology and software by utilizing cloud infrastructure. Thanks to this cost-effectiveness, businesses can invest more in innovation and cutting-edge Industry 5.0 technologies.
f. **Security and Compliance**: To safeguard sensitive data and guarantee compliance with industry laws, cloud providers provide strong security features and compliance solutions. In Industry 5.0, where data security and privacy are critical, this is essential.
g. **Flexibility and Innovation**: New technologies may be quickly tested and implemented thanks to the cloud. Without having to make large upfront investments, businesses can quickly embrace and test new Industry 5.0 developments, creating an environment that is more inventive and nimble.

12.2.2 Blockchain

Industry 5.0 can be greatly improved by blockchain technology since it offers efficient, transparent, and safe methods of managing transactions and data [6–10]. How blockchain helps Industry 5.0 technology is as follows:

a. **Enhanced Security**: Data is protected against alteration and illegal access thanks to the blockchain's decentralized and cryptographic structure. Blockchain contributes to data integrity and fraud prevention in Industry 5.0, where sensitive data from IoT devices, manufacturing processes, and AI systems are regularly exchanged.
b. **Transparency and Traceability**: Blockchain technology offers an unchangeable ledger that records each transaction and modification and prevents them from being changed in the past. Ensuring compliance with norms and standards and tracking the lifespan of products from raw ingredients to completed items depends heavily on transparency.
c. **Supply Chain Management**: By offering a single, shared version of the truth, blockchain technology can enhance supply chain visibility. This aids in monitoring the source of resources, confirming the legitimacy of goods, and cutting down on fraud and inefficiencies in the supply chain.
d. **Data Integrity**: Blockchain guarantees the accuracy and immutability of data entered into the system. Maintaining the quality and dependability of data is essential for AI-driven decision-making, predictive maintenance, and sophisticated manufacturing processes.
e. **Decentralization**: Blockchain runs on a distributed network of nodes, as opposed to conventional centralized databases. By ensuring that

data and transactions are resistant to intrusions and system failures, this decentralization lowers the possibility of single points of failure.

f. **Authentication and Identity Management**: Blockchain can improve security and expedite procedures for confirming an individual's identity. Blockchain technology can effectively handle digital identities and verify users and devices in Industry 5.0, where various stakeholders are involved.

12.2.3 Big data analytics

Big data analytics is essential to Industry 5.0 because it uses vast amounts of data to improve operational effectiveness, inform decisions, and develop more customized and sophisticated production processes. Big data analytics helps Industry 5.0 technologies in the following ways:

a. **Predictive Maintenance**: By analyzing enormous volumes of data from machines and IoT sensors, big data analytics can foresee equipment faults before they happen. Predictive maintenance helps minimize downtime, optimize maintenance schedules, and prolong the life of machinery by evaluating previous performance and real-time data.
b. **Quality Control**: Big data analytics can be used to find trends and abnormalities that impact the quality of products by analyzing data gathered from different manufacturing stages. This enables modifications and enhancements in real time, guaranteeing that goods fulfill strict requirements and cutting down on waste.
c. **Process Optimization**: By offering insights into manufacturing processes, big data analytics helps manufacturers find inefficiencies and streamline operations. Businesses may increase productivity, optimize processes, and save energy costs by evaluating data from sensors, supply networks, and production lines.
d. **Customization and Personalization**: Industry 5.0 places a strong emphasis on producing individualized goods. By analyzing consumer preferences, habits, and feedback, big data analytics enables businesses to create personalized goods that live up to individual needs and expectations.
e. **Supply Chain Management**: Better demand forecasting, inventory control, and logistical optimization are all made possible by analyzing data from all points of the supply chain. Real-time tracking and improved supplier collaboration are made possible by big data analytics, which lowers interruptions and boosts productivity.

12.2.4 6G and beyond

Industry 5.0 is expected to undergo significant changes due to the upgraded connectivity, ultralow latency, and improved data capacity offered by 6G and beyond [10–13]. These next-generation wireless technologies help Industry 5.0 in the following ways:

a. **Ultralow Latency**
   - **6G Capabilities:** Offers incredibly low latency, maybe in the microsecond range.
   - **Encouragement of Industry 5.0:** Permits real-time communication between humans, machines, and sensors. This is critical for applications like synchronized human–machine cooperation, real-time quality control, and autonomous robotics that need instant response.

b. **High-Speed Data Transfer**
   - **6G Capabilities:** Provides bandwidth and data speeds that are much higher than those of preceding generations.
   - **In Favor of Industry 5.0:** Allows for the quick transport of enormous amounts of data, which is necessary for applications like edge computing's (EC's) complicated AI models, real-time data analytics, and high-definition video surveillance.

c. **Massive Device Connectivity**
   - **6G Capabilities:** A large number of connected devices can be supported per square kilometer.
   - **Support for Industry 5.0:** Guarantees the smooth integration of a wide range of sensors, IoT devices, and intelligent machinery in production settings. Across the whole industrial ecosystem, complete data gathering, monitoring, and control are made possible by this interconnectedness.

d. **Advanced Edge Computing**
   - Using EC, 6G capabilities process data closer to the source.
   - **Encouragement of Industry 5.0:** Permits real-time data analysis at the edge, therefore lowering latency and bandwidth consumption. Applications like autonomous decision-making, predictive maintenance, and real-time process optimization depend on this.

e. **Enhanced Security**
   - **6G Capabilities:** Contains cutting-edge security mechanisms and features.
   - **Support for Industry 5.0:** Uses strong encryption and security methods to shield confidential industrial data from online attacks. By doing this, devices can communicate securely while preserving their operational integrity and intellectual property.

f. **Energy Efficiency**
   - **6G Capabilities:** Focuses on network operations and energy-efficient technology.
   - **Support for Industry 5.0:** Lowers operating expenses and promotes sustainable production practices by reducing the energy consumption of communication networks.

g. **Global and Reliable Coverage**
    - **6G Capabilities:** Provides better coverage and connectivity in hard-to-reach or difficult-to-reach areas.
    - **Encouragement of Industry 5.0:** Enables seamless operations and global data integration by guaranteeing consistent and dependable data communication across geographically scattered and remote industrial locations.

12.2.5 Internet of Things (IoT)

Industry 5.0 relies heavily on the IoT to link disparate systems and devices to facilitate smarter, more effective, and human-centered manufacturing processes. IoT helps Industry 5.0 technologies in the following ways:

a. **Real-Time Data Collection**
    - **IoT Capabilities:** Real-time data from machines, manufacturing lines, and environments is collected by sensors and devices.
    - **Encouragement of Industry 5.0:** Gives constant insight into manufacturing processes, allowing for prompt modifications and guaranteeing that operations are optimized by the most recent data.

b. **Predictive Maintenance**
    - **IoT Capabilities:** IoT sensors keep an eye on the functionality and health of equipment.
    - **Encouragement of Industry 5.0:** Uses data analysis to forecast possible equipment malfunctions before they happen. This minimizes maintenance expenses, increases the lifespan of machinery, and decreases unscheduled downtime.

c. **Process Optimization**
    - **IoT Capabilities:** Devices and sensors monitor a range of parameters about energy consumption, production efficiency, and quality.
    - **Encouragement of Industry 5.0:** Makes use of data to find inefficiencies, streamline operations, and improve production as a whole. Better resource usage, decreased waste, and more productivity result from this.

d. **Enhanced Quality Control**
    - **IoT Capabilities:** Throughout the production process, monitors and logs quality metrics.
    - **Encouragement of Industry 5.0:** Gives instant access to information on the quality of products, allowing for quick correction and guaranteeing that goods live up to expectations.

e. **Personalization and Customization**
    - **IoT Capabilities:** Compiles information on manufacturing needs and consumer preferences.
    - **Support for Industry 5.0:** Makes it possible to create personalized goods using data collected in real time and input from customers.

More adaptable production procedures that can meet specific requirements are made possible by this.

f. **Supply Chain Management**
   - **IoT Capabilities:** Keeps tabs on supply chain, logistics, and inventory activities.
   - **Encouragement of Industry 5.0:** Improves inventory control, demand forecasting, and supply chain visibility and coordination while lowering disruptions.

g. **Improved Human–Machine Interaction**
   - **IoT Capabilities:** Uses real-time data and user-friendly interfaces to link human operators with equipment and systems.
   - **Encouragement of Industry 5.0:** Improves decision-making, makes it easier for humans and machines to work together, and makes sure that operators can communicate with and operate systems more effectively.

h. **Remote Monitoring and Control**
   - **IoT Capabilities:** Makes it possible to access devices and systems remotely.
   - **Support for Industry 5.0:** Enables remote manufacturing process monitoring and control, which is especially helpful for managing operations across several locations and addressing problems without the need for on-site presence.

i. **Integration with AI and Machine Learning**
   - **IoT Capabilities:** Offers data that machine learning (ML) and AI algorithms can use.
   - **Encouragement of Industry 5.0:** Provides real-time data to algorithms, enhancing AI-driven insights and automation and resulting in more informed decisions, predictive analytics, and streamlined procedures.

12.3 CHALLENGES OF INDUSTRY 5.0

Industry 5.0, which blends cutting-edge technologies and prioritizes human-centric methods, presents several challenges in addition to the potential for considerable breakthroughs in manufacturing and industrial processes. The following are some of the main challenges that Industry 5.0 faces as reflected in Figure 12.2.

12.3.1 Integration complexity

- **Challenge:** Including cutting-edge technology like robotics, IoT, and AI in current systems can be expensive and time-consuming.
- **Impact:** Companies would have trouble attaining seamless interoperability and might have to make significant infrastructure upgrades.

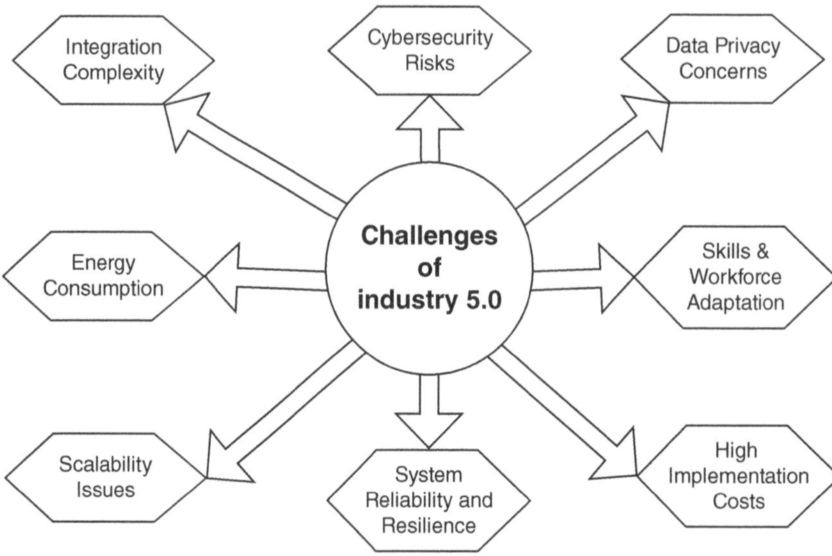

Figure 12.2 Industry 5.0 challenges.

12.3.2 Cybersecurity risks

- **Challenge:** Systems are more vulnerable to cybersecurity threats because of Industry 5.0's enhanced connection and data interchange.
- **Impact:** Strong security measures and ongoing attention are necessary due to the increased risk of cyberattacks, data breaches, and industrial espionage.

12.3.3 Data privacy concerns

- **Challenge:** The collection and processing of massive volumes of data raises questions about data privacy and regulatory compliance.
- **Impact:** By navigating the complex data protection standards, organizations must ensure that sensitive and personal data is handled securely and morally.

12.3.4 Skills and workforce adaptation

- **Challenge:** The workforce needs new skills and abilities as a result of the shift to more advanced technologies.
- **Impact:** Businesses may need to close a skills gap by retraining employees, hiring people with experience in developing technology, and other means.

12.3.5 High implementation costs

- **Challenge**: Investing a substantial amount of capital in new machinery, systems, and technologies can be necessary to implement Industry 5.0 technology.
- **Impact**: Small and medium-sized businesses can find it difficult to afford the cost of implementing these cutting-edge solutions.

12.3.6 System reliability and resilience

- **Challenge**: Keeping intricate, interconnected systems dependable and robust in various scenarios is a challenge.
- **Impact**: To avoid having a significant negative influence on operations, systems must be built to elegantly accept errors and interruptions.

12.3.7 Scalability issues

- **Challenge**: It can be difficult to scale Industry 5.0 solutions across many activities and geographical areas.
- **Impact**: In order to meet the diverse demands of various industries and business sizes, solutions must be flexible and scalable.

12.3.8 Energy consumption

- **Challenge**: Enhanced connectivity and cutting-edge technologies may result in higher energy usage.
- **Impact**: To promote sustainability goals, businesses must take the environment into account and look for energy-efficient solutions.

12.4 INDUSTRY 5.0 CREATIVE APPLICATIONS

12.4.1 Intelligent healthcare

ML models are being used by doctors these days to aid in patient diagnosis. This aids in increasing the precision of disease diagnosis, saving the patient a great deal of time and money. But given the circumstances, this is insufficient. The world urgently needs technology that can guarantee individualized patient requirements, such as tracking blood pressure and blood sugar levels, and providing patients with individualized treatment with medical help. This can be made feasible via Industry 5.0. Smart watches, intelligent sensors, and other wearable technology can continuously record a patient's medical history in real time, and this information can be saved on the cloud. The patients' medical conditions can subsequently be diagnosed using ML algorithms. These smart gadgets can interact with one another. Should a medical professional be called upon, these devices can relay the patient's current condition and notify the physician to attend to the patient. By using cobots, doctors can perform surgery on patients with the assistance of robots that can converse with one another. These are just a few instances of how Industry 5.0 could transform the healthcare sector [14–16]. This revolution facilitates the production of implants, customized gadgets, etc. By Industry 5.0, everyday tasks such as doctor check-ups can be handled by cobots. Doctors can focus on higher-level work in this way.

12.4.2 Cloud manufacturing

A unique approach to transforming the conventional manufacturing paradigm into an advanced manufacturing process is cloud manufacturing, which incorporates the newest technologies including virtualization, cloud and EC, IoT, and service-oriented technologies. Multinational stakeholders will work together to run an economical and effective manufacturing process in a cloud manufacturing setup. Reliability, excellent quality, cost-effectiveness, and on-demand capabilities are what set cloud manufacturing apart. Additionally, it benefits the environment because cloud manufacturing can do away with the need for raw materials to be sent across great distances for the manufacturing process. The next generation of cloud manufacturing systems, known as Industry 5.0, is anticipated to support increasingly complex and varied needs in the contexts of engineering, production, and logistics. The development of EC features, 5G-based telecommunication networks, and AI/ML technologies opens up new ways to greatly increase the capabilities of cloud manufacturing systems in the future [17,18].

12.4.3 Supply chain management

Disruptive technologies driving Industry 5.0—such as collaborative robots (cobots), machine learning, the Internet of Things, edge computing, 5G and beyond, digital twins, and others—when integrated with human intelligence and creativity, enable companies to meet demand more efficiently and deliver highly personalized and customized products at a faster pace.

12.4.4 Manufacturing/production

Industry 5.0 is the next stage of industrial development, emphasizing human–machine cooperation to create more efficient, sustainable, and customized manufacturing processes. In contrast to Industry 4.0, which placed a strong emphasis on automation and data interchange via IoT and AI technologies, Industry 5.0 restores human-centric innovation to the forefront. In the manufacturing sector, this entails fusing human ingenuity and problem-solving abilities with sophisticated robotics and AI. As a result, production becomes more customized, with humans supervising and optimizing processes to ensure flexibility and adaptability while machines take care of repetitive chores. Furthermore, Industry 5.0 emphasizes social responsibility and sustainability heavily, pushing manufacturers to embrace eco-friendly procedures and make constructive contributions to society [19,20].

12.4.5 Smart education

Industry 5.0, which combines cutting-edge technology with a human-centered approach to learning, has the potential to completely transform intelligent education. In contrast to other industrial revolutions that prioritized technology alone, Industry 5.0 highlights how AI and human intelligence can work together to improve personalized learning. This applies to AI-driven teaching technologies in smart education that adjust to each student's unique learning preferences and requirements, delivering personalized lesson plans and immediate feedback. While AI takes care of administrative duties and data analysis, instructors can employ these tools to concentrate more on emotional intelligence, creativity, and critical thinking. Industry 5.0 also promotes sustainability and ethical technology use by encouraging the integration of social and ethical issues into schooling. Industry 5.0 seeks to build a more flexible, inclusive, and engaging learning environment that equips students for the challenges of the future by fusing technical innovation with human understanding.

12.4.6 Disaster management

By combining cutting-edge technology with human knowledge to improve preparedness, response, and recovery activities, Industry 5.0 revolutionizes disaster management. These days, the emphasis is on developing intelligent systems that collaborate with people rather than just automating activities. Real-time data analysis and early warning systems are made possible by the increased accuracy with which robots and AI-driven tools can predict and monitor crises. Autonomous vehicles and drones can help with search and rescue efforts by entering places that might be too hazardous for people. Furthermore, Industry 5.0 highlights the significance of sustainability and social responsibility, promoting the creation of environmentally friendly solutions and resilient infrastructure that lessen the effects of disasters on communities. Industry 5.0 improves disaster management techniques by fusing technical innovations with human intuition and creativity.

This makes disaster management tactics more effective, flexible, and adaptable to changing circumstances.

12.5 POTENTIAL DIRECTIONS FOR FUTURE RESEARCH

Prospective areas for further research on Industry 5.0 could include extending its impact across multiple sectors and expanding our understanding in several areas:

12.5.1 Human–machine collaboration

Looking at the seamless integration of robotics, advanced AI, and human intelligence across several industries. This involves researching the most effective methods of combining mechanical efficiency and human creativity and problem-solving to maximize results.

12.5.2 Ethical and social implications

Investigating the socioeconomic effects and moral dilemmas raised by Industry 5.0 technology. Research may concentrate on how Industry 5.0 may affect employment, privacy, and data security, as well as how to make sure that the advantages are shared fairly throughout society.

12.5.3 Sustainability and green technologies

Analyzing the potential contributions of Industry 5.0 to environmental sustainability. This entails creating environmentally friendly production methods, cutting waste, and investigating how circular economy ideas apply to smart manufacturing and other industries.

12.5.4 Personalization and customization

Looking into how Industry 5.0 technology can make it possible for goods and services to be hyperpersonalized. This study can concentrate on sectors where customized solutions can significantly enhance user experiences and results, such as healthcare, education, and retail.

12.5.5 Resilience and adaptation

Investigating how Industry 5.0 may improve supply chains, healthcare, and disaster management, among other vital systems, in terms of resilience. The goal of research can be to create flexible systems that require little human intervention to respond to unforeseen obstacles and interruptions.

12.5.6 Interdisciplinary applications

Increasing the amount of research being done on the potential applications of Industry 5.0 in smart cities, agriculture, and energy, among other domains. Recognizing how various industries may use human–machine cooperation to solve difficult issues and improve innovation.

12.5.7 Skill development and workforce transition

To ensure that workers can adapt to new technology and prosper in a changing labor market, research can examine educational initiatives, training programs, and legislative interventions.

12.6 CONCLUSION

Industry 5.0, which combines human-centric approaches with cutting-edge technology like AI, robotics, and IoT, offers several potential. This improves sustainability, encourages creativity, and customizes output. Nonetheless, obstacles encompass moral dilemmas, changes in the workforce, and the requirement for strong cybersecurity. Attention must also be paid to the challenge of merging human and machine collaboration. Subsequent investigations ought to concentrate on tackling these obstacles, examining the possibilities of Industry 5.0 in other domains, and formulating structures that guarantee fair and enduring advantages for industries and the community as a whole. Industry 5.0 can realize its full potential by tackling these issues.

REFERENCES

1. Varela-Vaca, Á., et al. "Towards the Realization of Industry 5.0: Human-Centric AI for Sustainable Manufacturing." *IEEE Access*, vol. 11, 2023, pp. 36578–36593.
2. Zhang, Y., & Li, W. "Industry 5.0: Evolution, Impacts, and Future Prospects." *IEEE Trans. Eng. Manage*, vol. 70, no. 3, 2023, pp. 1016–1029.
3. John, S., & Mehta, A. "Sustainable Innovation in Industry 5.0: Integration of Digital Twins and Circular Economy." *IEEE Trans. Ind. Inf.*, vol. 19, no. 2, 2023, pp. 854–863.
4. Ahmed, A., & Chen, Z. "The Role of Artificial Intelligence in Industry 5.0: Opportunities and Challenges." *IEEE Trans. SMC: Syst.*, vol. 53, no. 4, 2023, pp. 732–740.
5. Sharma, R., & Gupta, S. "Human-Centric Smart Factories: Bridging the Gap between Industry 4.0 and Industry 5.0." *IEEE IoT J.*, vol. 10, no. 7, 2023, pp. 4876–4885.

6. Na, X., & Wang, Y. “Steps toward Industry 5.0: Building ‘6S’ Parallel Industries With Cyber-Physical-Social Intelligence.” *IEEE/CAA J. Autom. Sin.*, vol. 10, no. 6, 2023, pp. 2390–2402.
7. Wang, F.-Y., & Yang, J. “Industry 5.0: From Cyber-Physical Systems to Cyber-Physical-Social Systems.” *IEEE Intell. Syst.*, vol. 38, no. 2, 2023, pp. 32–45. doi:10.1109/MIS.2023.3247520.
8. Kaur, J., & Singh, S. “Industry 5.0: Leveraging Human-Machine Collaboration for Enhanced Productivity.” *IEEE Trans. Autom. Sci. Eng.*, vol. 20, no. 4, 2023, pp. 1515–1525.
9. Chen, H., & Liu, Y. “Exploring Industry 5.0’s Potential in Autonomous Manufacturing Systems.” *IEEE Robot. Autom. Lett.*, vol. 8, no. 1, 2023, pp. 65–73.
10. Arora, K., Joshi, G. P., Ragab, M., Rawa, M., Milyani, A. H., Mansour, R. F., & Yang, E. “Bilateral Contract for Load Frequency and Renewable Energy Sources Using Advanced Controller.” *Comput. Mater. Contin.*, vol. 73, no. 2, 2022, pp. 3165–3180. https://doi.org/10.32604/cmc.2022.026966.
11. Morales, R., & Chen, X. “Challenges and Opportunities in the Transition to Industry 5.0: A Case Study in the Automotive Sector.” *IEEE Access*, vol. 11, 2023, pp. 12345–12359. doi:10.1109/ACCESS.2023.3280912.
12. Gupta, P., & Srivastava, R. “AI-Driven Supply Chains: Industry 5.0 Perspectives.” *IEEE Trans. Eng. Manag.*, vol. 70, no. 4, 2023, pp. 890–899. doi:10.1109/TEM.2023.3257894.
13. Shukla, P., & Patel, R. “Towards Green and Human-Centric Manufacturing: The Role of Industry 5.0.” *IEEE Trans. Indus. Electron.*, vol. 71, no. 3, 2023, pp. 1456–1464. doi:10.1109/TIE.2023.3245782.
14. Arora, K., Kumar, A., Kamboj, V. K., Prashar, D., Jha, S., Shrestha, B., & Joshi, G. P. “Optimization Methodologies and Testing on Standard Benchmark Functions of Load Frequency Control for Interconnected Multi-Area Power System in Smart Grids.” *Mathematics*, vol. 8, no. 6, 2020, p. 980. https://doi.org/10.3390/math8060980.
15. Kumar, A., & Verma, P. “Cybersecurity Challenges in Industry 5.0: A Review.” *IEEE Secur. Privacy*, vol. 21, no. 2, 2023, pp. 30–38. doi:10.1109/MSEC.2023.3287456.
16. Ali, M., & Khan, N. “Industry 5.0 and the Circular Economy: A Path towards Sustainable Development.” *IEEE Access*, vol. 11, 2023, pp. 22356–22368. doi:10.1109/ACCESS.2023.3296712.
17. Peterson, D., & Moore, A. “AI-Augmented Robotics in Industry 5.0: Enhancing Human-Robot Collaboration.” *IEEE Trans. Robot.*, vol. 39, no. 1, 2023, pp. 175–188. doi:10.1109/TRO.2023.3289125.
18. Bhatt, B., Sharma, H., Arora, K., Joshi, G. P., & Shrestha, B. “Levy Flight-Based Improved Grey Wolf Optimization: A Solution for Various Engineering Problems.” *Mathematics*, vol. 11, no. 7, 2023, p. 1745. https://doi.org/10.3390/math11071745.
19. Rossi, M., & Bianchi, P. “Integrating Digital Twins with Industry 5.0: A New Paradigm in Manufacturing.” *IEEE Trans. Syst. Man Cybernet.* Syst., vol. 53, no. 5, 2023, pp. 1840–1850. doi:10.1109/TSMC.2023.3279104.
20. Lee, C., & Kim, H. “Industry 5.0: Synergies between Human and Artificial Intelligence.” *IEEE Trans. Neural Netw. Learn. Syst.* vol. 34, no. 3, 2023, pp. 960–970. doi:10.1109/TNNLS.2023.3285714.

Chapter 13

Simulation and analysis of battery state of charge (SOC) using MATLAB for energy management systems

Ranjit Kumar Bindal, Prashant Kumar, Vaibhav, and Vaibhav Jadon

13.1 INTRODUCTION

The introduction within the current study provides a venture to conduct a review on the accuracy of state of charge (SOC) approximation in battery management systems (BMS) and presents a MATLAB-based simulation platform to engineer and evaluate SOC when in charging and discharging modes of operation. SOC, as an important parameter, provides an indication of the remaining energy capacity of the battery as a percentage of its total capacity. The lack of accurate SOC estimates can lead to poor performance, inefficient operation, or unsafe operating conditions with rechargeable batteries that are why SOC estimates are important for reliable battery operation. The need for accurate SOC estimation is also an essential consideration in optimizing battery cycling, safety when recharging, battery longevity, and many applications where rechargeable batteries are in operation (e.g., electric vehicles, renewable energy storage schemes, and movable electronics). With regard to these applied scenarios, a battery normally undergoes various charge and discharge scenarios, thereby requiring an understanding of the pattern of how SOC varies with time to engage effective energy management. Despite the complexities of SOC estimation due to load conditions, temperature, and battery aging, this study used the coulomb counting method, which is a reliable common SOC estimation technique that is easy to implement (Figure 13.1).

This approach calculates the SOC by summing the net current entering or exiting the battery over time, providing an instantaneous approximation of remaining capacity in the battery. The MATLAB simulation developed in this study provides a realistic and user-friendly platform for simulating SOC behavior under controlled conditions. By specifying significant parameters such as battery capacity, initial SOC, charging current, discharging current, and time step, the simulation replicates the SOC behavior for a given period of time and delivers helpful information about the way the battery behaves under different conditions of operation. This strategy works by tracking

DOI: 10.1201/9781003641582-13

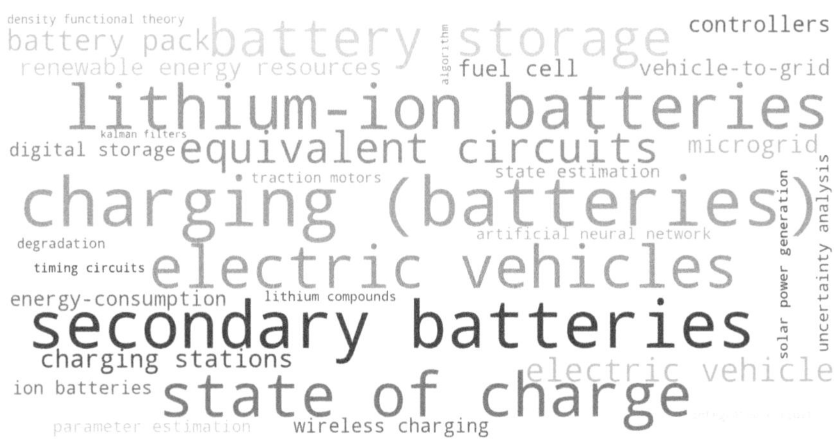

Figure 13.1 Some important keywords.

the current that flows into and out of the battery over a specified duration, integrating the current to ascertain the net charge added or withdrawn. As with any approach, it is simple and widely applicable, yet the technique's precision depends on numerous aspects, including sensor accuracy, self-discharge compensation, and SOC calibration drifting. Because of this, the MATLAB-bound simulation environment was deliberately constructed to consider these factors and scrutinize their impact on SOC estimation accuracy. In the context of practical BMS applications, SOC estimation must be sensitive to and track real-time alterations due to rapid and persistent changes in power demand, regenerative braking, or intermittent charging like those found in electric vehicles. The simulation not only incorporated constant load profiles but also included step current functions, changes in ambient temperature, and parasitic losses to capture real-world usage scenarios. This provided a better assessment of SOC behavior across different operational modes. Additionally, MATLAB was chosen due to its immense computational power for complex analysis, graphical representation capabilities, and advanced modeling features through Simulink, which enables the development of both elementary and sophisticated battery models. To increase the relevance of the analysis with the prevailing industry practices, the primary battery chemistry simulated was lithium ion, which is widely adopted in EVs and portable electronics owing to their high energy density and prolonged life cycle. This study is focused on understanding a literature review for SOC estimation methods beyond coulomb counting. Some of the methods include open-circuit voltage (OCV), Kalman filtering, neural networks, fuzzy logic, and hybrid estimation methods. Each SOC estimation method offers unique benefits and disadvantages in relation to precision, estimation, noise, temperature, and complexity. Even though hybridized methods are becoming more advanced, coulomb counting is often still used in embedded systems because it is straightforward and requires minimal computational resources. To address the weaknesses of coulomb counting, methods such as periodic SOC calibration using OCV or temperature correction models can be applied. It is also important to note that the simulation includes SOC estimation methods and the degradation of battery lifespan as a SOC estimating parameter. Increasing age results in elevated internal resistance and decreased capacity of a battery. If these parameters are not considered, they may result in SOC calculations that are close to incorrect. The study depicts regular recalibration and adaptive SOC slope estimation algorithms as solutions to these issues. Also important is the inclusion of over-charging and deep discharging protective methods, which help prevent damage to the battery while ensuring operational safety. To ensure that the results generated from the simulation are both educative and helpful, it has been resolved that the results are graphed, demonstrating how SOC changes with time for different loads and temperatures. Such graphs can be used for determining any trends, outliers, or other ranges, where estimation errors might be possible to occur. At the end, the simulation can be a starting point in order to develop and verify advanced BMS designs based on deep learning algorithms for predictive SOC estimation and automatic control algorithms. Owing attention to SOC estimation—key for effectiveness/safety /cost-efficiency of battery-operated systems—the study combines theoretical knowledge with hands-on experience toward more precise battery management. Such a kind of research not only enriches science but also remains an easy source of information for engineers and developers dealing with energy storage systems (ESS). In the end, SOC estimation with the right precision leads to efficient energy resources' management, especially regarding the persuaded tendency toward electricity auto-transport and decentralized renewable energy structures.

13.2 LITERATURE REVIEW

Zhang and Xiang [1] proposed a novel and high-accuracy SOC approximation strategy for lithium-ion batteries based on a dual optimization particle swarm sparrow search algorithm. Their method allows flexibility to track multitemperature conditions with improved accuracy and SOC estimates, which are key to battery efficiency and lifespan characteristics. Song et al. [2] investigated electric heavy truck battery performance concerning motor waste heat recovery and highlight how managing temperature can be used to maximize battery lifespan and mileage while proposing a sustainable method of maximizing energy for EVs. Wang et al. [3] proposed an SOC-dependent charge scheduling approach for optimizing heavy-duty electric truck battery exchange stations. The study offers low-cost energy management techniques with assured grid stability. Okonkwo et al. [4] formulated a mathematical model of lithium-ion polymer battery charge and discharge dynamics. The new model enhances the forecast of battery performance for different operating modes, which facilitates improved ESS design. Huang et al. [5] proposed a capacity optimization model of PV-battery storage systems in electric vehicle charging stations (EVCS). In their study, they undertake multivenue charging behavior with economic efficiency and grid stability as the most important ones. Wang et al. [6] discussed rapid SOC balance control techniques of segmental battery ESS. In their method, they optimize battery life by maintaining a uniform charge among battery modules. Bera et al. [7] introduced a novel SOC measurement method based on an extended Kalman filter (EKF) and fuzzy logic. Their effort improves SOC estimation precision through lower measurement noise and lower computational loads. Shen et al. [8] developed a low-communication controller framework for multiobjective energy management in direct current (DC) microgrids. Their effort is encompassed through incorporating battery storage facilities and demand side management to ensure optimal energy use. Gupta and Suhag [9] developed a hybrid energy framework by combining various battery and hydrogen recharging stations inside independent microgrids. Their system improves voltage regulation and energy customization, thus making energy management more responsive (Table 13.1).

Paulech and Bat'A [10] studied battery module thermal management systems in electric vehicles and determine the most efficient thermal management techniques. Their research helps enhance battery efficiency and increase lifespan through efficient heat removal. He et al. [11] introduced a sliding method spectator for SOC approximation of lithium-ion battery voltage dynamics. Their method improves the accuracy of estimation, especially under dynamic load conditions. Reddy and Reddy [12] explored fuel cell-powered EVCS, investigating in particular the possibility of bridging grid energy with renewable energy. Their study examines the efficiency and feasibility of hydrogen-based EV infrastructure. Arora et al. [13] contrasted and optimized energy management policies for hydrogen fuel cell electric vehicles. Their analysis plots the interplay of energy efficiency, cost, and environmental sustainability. Rekioua et al. [14] introduced an optimal control approach for hybrid energy storage-based wind turbine systems. Their system incorporates hybrid Maximum Power Point Tracking techniques for enhanced efficiency in energy harvesting. Lahmer et al. [15] presented a consensus-based adaptive distributed categorized control system for battery energy storage of a DC microgrid. Their article focuses on decentralized energy management issues as well as enhancing system robustness. Cao et al. [16] investigated battery energy storage schemes providing dynamic suppression frequency response services. Their article shows the significance of battery storage in ensuring grid stability and reliability. Han [17] proposed an optimal policy for charging current with a multistage constant current strategy utilizing a dandelion optimizer. Their protocol optimizes lithium-ion battery charging with lower degradation. Zhang et al. [18] presented a synthetic dataset for LG M50 battery degradation modes. Their

Table 13.1 Summary of references

Ref. no.	*Author(s) (year)*	*Title*	*Key findings*	*Research gaps*
[1]	Zhang and Xiang (2024)	High-precision SOC estimation using a dual optimization particle swarm sparrow search algorithm.	Improved SOC estimation under multitemperature situations, enhancing battery performance and longevity.	Needs real-world validation under diverse environmental conditions.
[2]	Song et al. (2024)	Effect of motor waste heat recovery on electric heavy truck batteries.	Demonstrates how thermal management improves battery life and driving range.	Limited testing on different EV models and real-time performance analysis.
[3]	Wang et al. (2024)	SOC-dependent charge scheduling for heavy-duty truck battery swap stations.	Optimizes battery swap station energy. Management while minimizing costs.	Requires further exploration of scalability and impact on power grids.
[4]	Okonkwo et al. (2024)	Mathematical prototypical of discharge and charge in polymer-based lithium-ion batteries.	Provides an improved mathematical framework for battery charge-discharge dynamics.	Needs experimental validation and incorporation of temperature variations.
[5]	Huang et al. (2024)	Capacity optimization of PV-battery storage for EVCS.	Enhances economic efficiency of battery storage in EVCS.	Further research needed on integration with smart grid systems.

dataset is extremely useful to machine learning-based degradation prediction models in the context of battery aging in various operating modes. Arora et al. [19] employed primary-principle calculations to explore Janus 2D-TiSSe as an anode electrode for lithium, sodium, and magnesium-ion batteries. Their research effort aids in the design of more advanced battery materials with a higher energy density. Tabine et al. [20] introduced a novel polynomial fitting procedure for SOC approximation in Li-ion batteries with temperature hysteresis. Their approach enhances SOC prediction accuracy, which helps improve battery management. Liu et al. [21] explained how chemo-mechanics at the grain level contribute to solid-state lithium battery cathode degradation. They present new material-level degradation mechanisms, which can be used to help reduce battery or cell deterioration mechanisms and improve battery performance and longevity.

13.3 METHODOLOGY

The approach for this research is a simulation in MATLAB for approximating the SOC of a battery during its charging and discharging. SOC is a major characteristic of BMS; SOC is defined as the battery capacity available articulated as a percentage of the total capacity of the battery. A straightforward approach called coulomb counting is used in the simulation, to compute SOC by adding the net current [in or out of the battery] per second. The simulation contains important parameters such as the capacity of the battery, the battery's initial SOC, the charge current, the discharge current, and the time step. In this simulation, the battery has a capacity of 10 ampere-hours (Ah) and an initial SOC of 50%. The charge current is 3 A; the discharge current is 2 A. The duration of the simulation has a maximum of 3,600 seconds (1 hour), with a time step of 1 second. The SOC is recomputed each time step based on the difference of charging and discharging current scaled by the capacity of the battery to give a percentage.

This method allows for a simple but effective identification of battery performance, charge management practices, and BMS design practices. The simulation can further encompass some of the more complex metrics such as temperature effects, battery life changing due to the rate of use, and sophisticated SOC estimations for better accuracy and ease of use in a physical application (Figure 13.2).

13.4 RESULT AND EVALUATION

The MATLAB simulation displays the SOC variation of the battery during 60 minutes starting from 50% against time. The simulation plot shows a direct rise in the SOC until it reaches its highest point of 62% at the 60-minute interval. The graph indicates SOC expansion to 62% during 60 minutes because of the defined 1 A charging combination of 3 A charging and 2 A discharging according to MATLAB code specifications. The range of SOC values lies inside the prescribed operating zone from 20% to 90%, thus preventing violations of genuine battery conditions and preventing unwanted overcharge or deep discharge occurrences. The straight SOC curve confirms the proper implementation of coulomb counting methodology that calculates SOC by integrating net current flows. The result confirms that the battery SOC progresses in direct proportion to net charge current during the charging process (Figure 13.3).

The simulation also reinforces the important task of setting proper SOC limits, such that the battery can be used safely and effectively. Overall, the simulation is capable of simulating battery SOC behavior under many conditions and gives a good rough idea of how some of the SOC will change with battery performance over time.

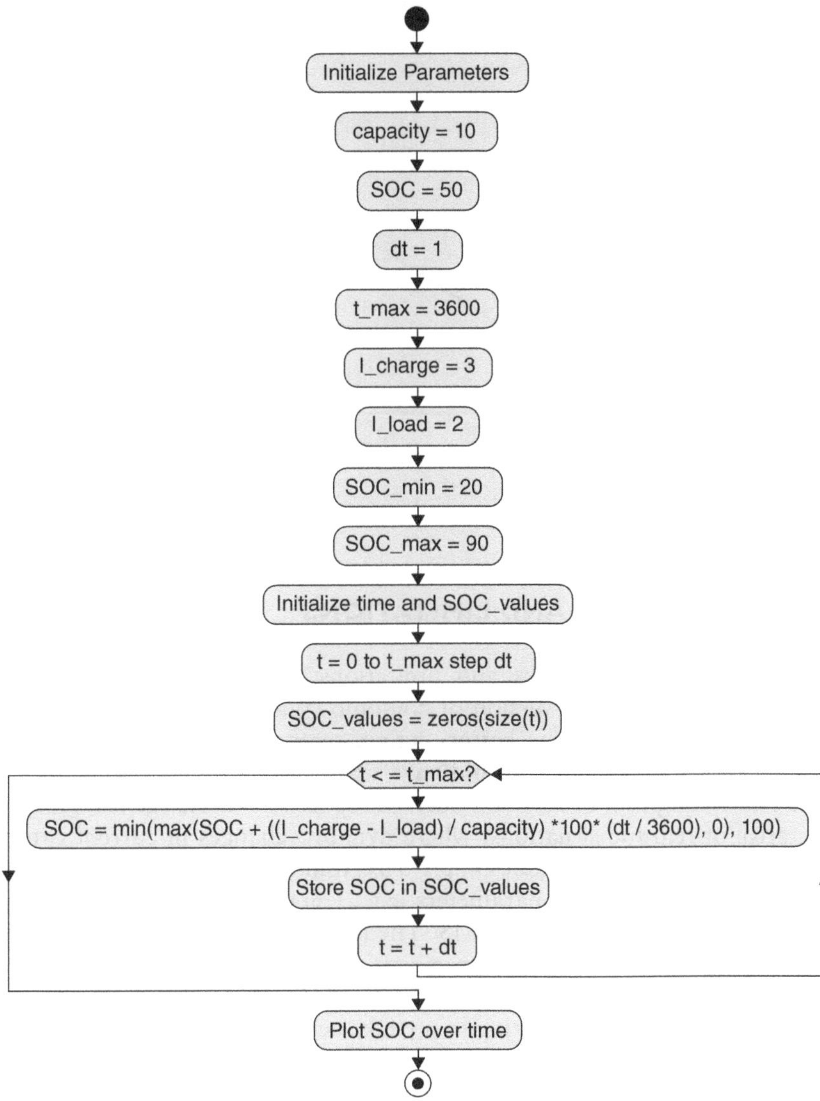

Figure 13.2 Proposed methodology.

This paper provides a good baseline to further research more advanced SOC estimations, thermal behavior, and battery aging, with possible future applications in electric vehicles, renewable energy storages, and other battery systems. The results show the applicability of the simulation toward optimizing battery performance and designing resilient BMS.

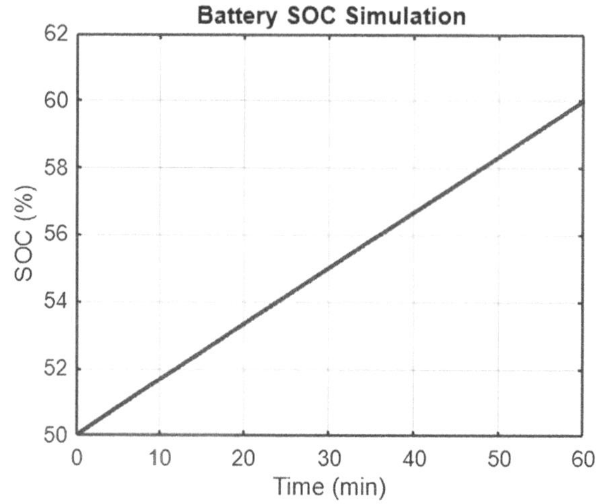

Figure 13.3 Battery SOC simulation.

13.5 CHALLENGES AND LIMITATIONS

There are some drawbacks and limitations to the simulation. First, it depends on a simplified version of coulomb counting and ignores sophisticated parameters such as battery aging, temperature effects, and internal resistance, all of which would have significant impacts on SOC in real life. The assumption of constant current charge and discharge is not representative of real-life situations, where dynamic changes in current can be expected. The simulation does apply SOC limits (20%–90%) but does not consider other important parameters (e.g., voltage limits and thermal limits) necessary for all-inclusive BMS. The model also did not consider noise or error in measurements, as these occur and impact SOC accuracy in real systems. Finally, the simulation is only for one battery cell and would incur other issues when extrapolating to a pack such as cell balancing and intercell differences. Additionally, actual battery operation is affected by a wide range of factors such as state of health, depth of discharge, environmental factors, and manufacturing tolerances, none of which are directly accounted for in the simulation. Battery aging specifically causes progressive capacity loss and rising internal impedance, which directly impacts SOC estimation accuracy over time. In real-world applications, algorithms need to be adaptive to counteract these degradations. Omitting such considerations in the simulation can provide a misleadingly positive image of SOC estimation accuracy and might result in unsatisfactory scalability when applying similar logic to real-time BMS implementations. Moreover, the exclusion of thermal dynamics is a serious shortcoming. Heat generation during high-rate discharging or charging can change battery performance, shift internal chemical equilibria, and introduce nonlinear effects not included in reduced models. A more accurate simulation would incorporate thermal coupling or even electrothermal modeling to represent such interactions. This is particularly relevant in the case of fast charging where temperature rise is not negligible and could potentially result in thermal runaway if not controlled well. Without considering thermal effects in a simulation, it loses its predictive power regarding temperature-related issues and thus underrepresents one of the principal dangers in battery systems. The other limitation is the accuracy of sensor measurements and precision. In the actual system, voltage and current sensors have drift, quantization error, or electromagnetic interference that may severely corrupt the SOC calculations. These were not included in the simulation, assuming ideal conditions. For a durable BMS, however, error mitigation techniques such as signal filtering, sensor calibration routines, or redundant sensing are utilized and should be addressed in advanced modeling endeavors. In addition, this simulation does not capture computational limitation or latency that can be encountered in real-time embedded applications. Real BMS hardware is limited in its processing power, memory, and energy resources, which puts constraints on SOC estimation algorithm complexity and execution time. While MATLAB provides a powerful and convenient environment for algorithm development and simulation, these models are commonly translated into hardware-implementable code with accommodations and optimizations not typically addressed at the simulation level. Cell-level simulation also neglects issues in multicell designs. In a pack of cells in a battery, cell imbalance through manufacturing variation, varying usage profiles, or temperature gradients creates dissimilar SOC states in the cells. Cell balancing methods—either passive (resistive shunting) or active (redistribution of charge)—become necessary for achieving equal performance and extending the lifespan of batteries. The lack of pack-level dynamics in the model restricts the model's generalizability to actual EV or ESS applications where having hundreds of cells under simultaneous management is the norm. Further, the model employs a perfect charging/discharge profile and does not implement regenerative braking effects or inconsistent charging interruption typical in automotive or renewable integration domains. All of these are characteristics that impact the dynamics of the SOC and to which the algorithm must react responsibly and stably under diverse conditions. Finally, the simulation does not have a validation step using experimental data. In real applications, it is necessary to validate SOC estimation against lab-measured or field data so as to confirm the reliability of the model and to determine its parameters. Without validation, the simulation outputs are hypothetical and will not necessarily reflect anticipated real-world performance. In spite of these constraints, the simulation offers a basic understanding of SOC estimation with a simple coulomb counting method. It is a stepping stone toward more detailed modeling, which would encompass electrochemical properties, dynamic environmental factors, and real-time requirements. In future research, adding pack-level attributes, incorporating real measurement data, and improving the fidelity of thermal and aging models would strengthen the simulation and make it useful in real BMS development.

13.6 FUTURE OUTCOMES

Future research on this study could include better simulation through the integration of progressive SOC estimation methods, such as Kalman filtering, machine learning, or neural networks, in a bid to achieve greater accuracy across different conditions. Adding temperature dynamics would allow the study of thermal effects on battery performance and SOC, and simulating aging of batteries would provide a more realistic SOC prediction over the lifespan of the battery by taking into account capacity fade and internal resistance changes. Additionally, replacing constant currents with real current profiles, e.g., from electric vehicles or renewable energy systems, would make simulations of operating conditions more realistic. Also, if we add a variable current load from the real-world driving cycles, such as Worldwide Harmonized Light Vehicles Test Procedure (WLTP) and Urban Dynamometer Driving Schedule (UDDS), it can be made more realistic and approximate simulations to real situations. These profiles represent the dynamic nature of electric vehicle use and renewable energy sources, where currents change widely on small time intervals. With this information, future simulations can quantify SOC estimation accuracy in complex operating environments and thus gain more insight into best practices in battery operation. As technology matures and the ESS become more complex, future studies have to focus on adaptive estimation algorithms that could respond to dynamic load profiles, dynamic changes in temperature, and aging effects in real time. Methods like the EKF, Unscented Kalman Filter, and Particle Filter have shown a lot of promise already in SOC tracking when the system dynamics are nonlinear and uncertain. The application of these algorithms to the model of the simulation may not only improve the accuracy but also lead to providing a more stable response to unexpected external forces, such as load transients, temperature fluctuations, and noise with measurements. Also, the artificial intelligence and machine learning models trained upon large empirical datasets could change the process of SOC estimation, allowing the system to recognize patterns, detect outliers, and adapt its behavior accordingly, which would allow the system to self-optimize its accuracy as a function of observed behavior. An important aspect of future work is electrothermal modeling as well wherein the relation between the electrical performance and the temperature is elaborately handled. This would facilitate thermal runaway risk prediction, improved cooling technique to be utilized, and optimized battery use—especially for fast-charge or high-discharge scenarios. Inclusion of such thermal dynamics would give design engineers the ability to validate safety margins in the simulation environment itself, drastically eliminating development risks and accelerating the validation cycles of products. Further, having a variable current load from an actual world driving like the WLTP or UDDS can introduce realism and approximate simulations to real environments. These profiles include the dynamic aspects of electric vehicle usage and renewable energy sources, where currents vary enormously in short periods of time. Using this information, upcoming simulations are able to determine the quality of SOC estimations in complex operations and thus provide further understanding of how to best manage batteries.

To further develop simulation scalability, upcoming research would also focus on modeling at the battery pack level, including series and parallel cell configurations, differences in intercell resistance, and balancing methods. Active balancing schemes, such as inductor or capacitor-based charge transfer, need to be simulated in order to test their effectiveness in improving system efficiency and SOC uniformity. This would further integrate the simulation model with real-world battery systems in electric vehicles, stationary grid energy storage, and industrial UPS applications. Ultimately, closing the gap between simulation and practical application needs to be the focus. This can be ensured via integration with experimental testbeds consisting of hardware-in-the-loop (HIL) setups or real-time digital simulators. By correlating simulated SOC estimates with actual battery hardware operated under diverse conditions, researchers will be able to establish confidence in the reliability and use of the model for real-world embedded BMS applications. Experimental correlation will also facilitate the refinement of model parameters, realization of measurement constraints, and evaluation of real-time computational tractability. Finally, the future work's roadmap is wide but necessary. By including advanced estimation algorithms, considering representative input profiles (temperature, aging), and modeling full-scale battery packs, we will be able to make SOC estimation tools more accurate, flexible, and pertinent to the swiftly evolving field of energy storage technologies.

13.7 CONCLUSION

This paper provided a MATLAB simulation of battery SOC based on coulomb counting. The simulation effectively illustrated the dynamic response of SOC during charging and discharging, where the SOC linearly increased over time as a result of a net charging current. The results confirmed the efficacy of the simulation in replicating battery operation within safe operating ranges (20%–90% SOC). Yet, the research also reported various shortcomings, such as the reduced model, current assumptions, and the absence of thermal and aging effects. Notwithstanding these shortcomings, the simulation presents a basic framework for subsequent research and development on BMS. Future research can explore the integration of sophisticated SOC estimation methods, thermal effects, aging, and dynamic current profiles to enhance the correctness and relevance of the model. This research contributes to the overall objective of maximizing battery performance, guaranteeing safe operation, and developing energy storage technology for uses like electric vehicles and renewable energy systems. For instance, variations in temperature will influence the internal resistance of the battery and indirectly modify the SOC behavior, so if they are not suitably compensated, wrong estimations result. Also, the capacity of the battery diminishing gradually because it ages will totally skew SOC readings over time when employing approaches based on a static nominal capacity like coulomb counting.

The inclusion of these components in upcoming simulation models can help fill the gap between theoretical calculation and real-world application. Specifically, the use of sophisticated algorithms such as the EKF, machine learning estimators, and data-driven predictive modeling can further enhance SOC estimation precision by addressing nonlinearities and system uncertainties. These methods can adjust to changing conditions and sensor noise, making them suitable for real-time BMS.

Additionally, scaling up the simulation from an individual cell to a multicell or entire battery pack setup would enable researchers to explore other issues like cell balancing, intercell variations, and the necessity for active or passive balancing techniques. Large battery pack dynamics, particularly in high-energy-demanding applications such as EVs and stationary ESS, are much more complex and require highly accurate control and estimation at both the pack and cell levels. More accurate modeling of these systems allows developers to create more durable BMS that prolong battery life, improve safety, and increase energy efficiency. Another area with high potential for future work is the incorporation of realistic charging and discharging profiles obtained from electric vehicle drive cycles or grid-connected storage systems. These profiles provide oscillating current curves, which are very different from the constant current sources in today's simulation. Simulation of such dynamic behavior would make the model far more relevant and useful to practical situations, providing greater insight into the real-world performance and difficulties of battery use under various load and environmental conditions. Finally, experimental verification of the enhanced simulation models by means of HIL testing or laboratory-based battery testing facilities would provide firm evidence of the accuracy of the simulation. This would also enable calibration of the model with respect to real-time data and hence make it suitable for direct deployment in embedded systems that are utilized in EVs, solar-storage devices, and consumer electronics. The use of accurate sensors and data acquisition hardware in these experiments would produce high-fidelity results, thereby contributing significantly to battery technology advancement.

REFERENCES

1. Y. Zhang and J. Xiang, "High-precision state of charge estimation of lithium-ion batteries based on dual optimization particle swarm sparrow search algorithm adaptive to multi-temperature conditions," *Journal of Energy Storage*, vol. 95, p. 112348, 2024, doi:10.1016/j.est.2024.112348.
2. Z. Song et al., "Analysis of the effect of motor waste heat recovery on the temperature and driving range of electric heavy truck batteries," *Batteries*, vol. 10, no. 9, p. 328, 2024, doi:10.3390/batteries10090328.
3. F. Y. Wang, Z. Chen, and Z. Hu, "Comprehensive optimization of electrical heavy-duty truck battery swap stations with a SOC-dependent charge scheduling method," *Energy*, vol. 308, p. 132773, 2024, doi:10.1016/j.energy.2024.132773.
4. F. N. Okonkwo, C. A. Okonkwo, and S. C. Nwigbo, "Mathematical model of discharge and charge in rechargeable polymer based lithium-ion battery," *Proceedings of the Institution of Mechanical Engineers, Part C: Journal of Mechanical Engineering Science*, vol. 238, no. 16, pp. 8244–8257, 2024, doi:10.1177/09544062241236097.
5. W. Huang et al., "Capacity optimization of PV and battery storage for EVCS with multi-venues charging behavior difference towards economic targets," *Energy*, vol. 313, p. 133833, 2024, doi:10.1016/j.energy.2024.133833.
6. J. Wang, S. Zhou, and J. Mao, "Research on fast SOC balance control of modular battery energy storage system," *Energies*, vol. 17, no. 23, p. 5907, 2024, doi:10.3390/en17235907.
7. C. Bera, R. Mandal, and A. Kumar, "A novel redacted extended Kalman filter and fuzzy logic-based technique for measurement of state-of-charge of lithium-ion battery," *Journal of Electrochemical Energy Conversion and Storage*, vol. 21, no. 4, p. 041003, 2024, doi:10.1115/1.4064096.
8. J. Shen, L. Wang, and L. Zhu, "A novel reduced-communication control strategy for multi-objective energy management of DC microgrids considering battery storage systems," *Journal of Energy Storage*, vol. 98, p. 112881, 2024, doi:10.1016/j.est.2024.112881.
9. A. Gupta and S. Suhag, "Hybrid structure integrating multiple battery and hydrogen charging stations in an autonomous microgrid for customized energy and voltage control," *Sustainable Materials and Technologies*, vol. 41, p. e01116, 2024, doi:10.1016/j.susmat.2024.e01116.
10. J. Paulech and M. Bat' A, "Analysis of battery module cooler for electric vehicles," *AIP Conference Proceedings*, vol. 3251, no. 1, p. 060005, 2024, doi:10.1063/5.0235322.
11. L. He et al., "State-of-charge estimation by extended sliding mode observer based on lithium-ion battery voltage dynamics," *Journal of Power Sources*, vol. 611, p. 234718, 2024, doi:10.1016/j.jpowsour.2024.234718.

12. B. P. K. Reddy and V. U. Reddy, "Analysis of fuel cell powered EV charging station," *Journal of Engineering and Technology for Industrial Applications*, vol. 10, no. 49, pp. 212–216, 2024, doi:10.5935/jetia.v10i49.1288.
13. Arora, K., Kumar, A., Kamboj, V. K., Prashar, D., Shrestha, B., & Joshi, G. P. (2021). "Impact of renewable energy sources into multi-area multi-source load frequency control of interrelated power system," *Mathematics*, vol. 9, no. 2, p. 186. https://doi.org/10.3390/math9020186.
14. D. Rekioua et al., "Effective optimal control of a wind turbine system with hybrid energy storage and hybrid MPPT approach," *Scientific Reports*, vol. 14, no. 1, p. 30013, 2024, doi:10.1038/s41598-024-78847-9.
15. A. Lahmer et al., "Consensus-based adaptive distributed hierarchical control of battery energy storage systems in a DC microgrid," *Journal of Energy Storage*, vol. 97, p. 112948, 2024, doi:10.1016/j.est.2024.112948.
16. X. Cao et al., "Battery energy storage systems providing dynamic containment frequency response service," *International Journal of Electrical Power & Energy Systems*, vol. 162, p. 110288, 2024, doi:10.1016/j.ijepes.2024.110288.
17. S. Han, "Optimal charging current protocol with multi-stage constant current using dandelion optimizer for time-domain modeled lithium-ion batteries," *Applied Sciences*, vol. 14, no. 23, p. 11320, 2024, doi:10.3390/app142311320.
18. H. Zhang, F. Altaf, and T. Wik, "Synthetic dataset of LG M50 batteries with different degradation pathways," *Data in Brief*, vol. 57, p. 111076, 2024, doi:10.1016/j.dib.2024.111076.
19. Arora, K., Joshi, G. P., Ragab, M., Rawa, M., Milyani, A. H., Mansour, R. F., & Yang, E. (2022). "Bilateral contract for load frequency and renewable energy sources using advanced controller," *Computers, Materials & Continua*, vol. 73, no. 2, 3165–3180. https://doi.org/10.32604/cmc.2022.026966.
20. A. Tabine et al., "A novel fitting polynomial approach for an accurate SOC estimation in Li-ion batteries considering temperature hysteresis," *e-Prime – Advances in Electrical Engineering, Electronics and Energy*, vol. 10, p. 100822, 2024, doi:10.1016/j.prime.2024.100822.
21. C. Liu, F. Roters, and D. Raabe, "Role of grain-level chemo- mechanics in composite cathode degradation of solid-state lithium batteries," *Nature Communications*, vol. 15, no. 1, p. 7970, 2024, doi:10.1038/s41467-024-52123-w.

Index